Rainbows of Rock, Tables of Stone

The Natural Arches and Pillars of Ohio

Frontispiece. The photogenic west entrance of Rock House, Ohio's largest natural arch and longest natural tunnel.

Rainbows of Rock, Tables of Stone

The Natural Arches and Pillars of Ohio

by

Timothy A. Snyder

The McDonald & Woodward Publishing Cor
Granville, Ohio

The McDonald & Woodward Publishing Company
Granville, Ohio

Rainbows of Rock, Tables of Stone
The Natural Arches and Pillars of Ohio

© 2009 by Timothy A. Snyder

Printed in Canada
by Friesens, Altona, Manitoba

First printing May 2009

16 15 14 13 12 11 10 09
1 2 3 4 5 6 7 8 9 10

Mixed Sources

Cert no. SW-COC-001271
© 1996 FSC

FSC

Library of Congress Cataloging-in-Publication Data

Snyder, Timothy A., 1948-
  Rainbows of rock, tables of stone : the natural arches and pillars of Ohio
/ by Timothy A. Snyder.
      p. cm.
  Includes bibliographical references and index.
  ISBN 978-0-939923-73-1 (pbk. : alk. paper)
  1. Natural bridges—Ohio. 2. Benches (Geomorphology)—Ohio. 3. Hoodoos
(Geomorphology)—Ohio. 4. Landforms—Ohio. 5. Geomorphology—Ohio. I.
Title.
  GB565.O3S69 2008
  551.4109771—dc22

                                                          2008038241

Contents

Preface vii

Introduction 1

Chapter 1
Conditions Required for the Formation
of Natural Arches and Pillars 15

Chapter 2
About Natural Arches 39

Chapter 3
A Catalog of Ohio's Natural Arches 81

Chapter 4
About Natural Pillars 269

Chapter 5
A Catalog of Ohio's Natural Pillars 289

Chapter 6
Fallen Arches and Toppled Tables 335

Chapter 7
Publicly Accessible Natural Arches and Pillars of Ohio 349

Epilogue 391

Appendixes
  I A Tabulated List of Ohio's Documented Natural Arches .. 394
  II A Tabulated List of Ohio's Documented Natural Pillars ... 402
  III English-Metric Conversion Tables 404

Bibliography 405

Index 419

To my father
Frank Wesley Snyder (1923–2001),
family camper and scoutmaster *extraordinaire*

Thanks for taking the time
to introduce me to the natural world.

Preface

It seemed like a good idea at the time. While attending the 1983 Natural Areas Association meeting held at Natural Bridge State Park in Kentucky, I joined a group making the steep climb up to view the feature for which the park was named. Although I had read of natural bridges, I had never actually seen one, and this particular example proved to be a spectacular introduction.

On the way down, I pondered aloud why Ohio had no such geological features. Mark Howes, a fellow District Manager with the Ohio Division of Natural Areas and Preserves and my hiking partner of the moment, replied that we did. As a matter of fact, he happened to be responsible for managing two of the largest: Rockbridge and Ladd Natural Bridge. This was a pleasant surprise, but "Only two?" I replied. "Don't we have rocks of similar kind and age as those here in Red Rock Gorge? Why does Kentucky have so many natural bridges while Ohio has only two?"

A few weeks later, Mark called. "Guess what I've found," he said. What he had found was Conkles Hollow Arch hiding in the depths of one of the state's most popular nature preserves and an area he, as manager, had explored many times. So Ohio had more than two natural arches after all! If this admittedly small example could have escaped notice in such a well-known area, how many more arches were lost in the hinterlands?

Mark's discovery prompted a flood of half-remembered images of arches I had run across in old books and odd sources of information over the years. Suddenly faced with a new area of interest, my enthusiasm swelled to accept it. Here was an opportunity to make a real contribution to the study of Ohio's natural history. I would make a list of all of Ohio's natural arches and visit each one where possible, measuring, mapping and photographing them to create a complete record of these unusual geologic features which apparently had been largely overlooked. It would not take long. After all, how many of these interesting features could remain undiscovered in this highly urbanized and intensively farmed state? We might find twelve or maybe twenty if we were really fortunate. Two summers of concentrated field work in my off hours should be adequate time to complete the job.

A quarter of a century later, the number of natural arches on my list surpassed eighty with more reported, but not yet visited. Once word of my little project got out, friends, relatives and complete strangers contacted me with potential leads. Some of them led to dead ends, like the report of a "natural stone bridge" in Cincinnati's Spring Grove Cemetery which turned out to be a road bridge built of glacial cobbles. Others, however, led to true gems like Castlegate Arch. An especially important discovery was the "Natural Bridge" folder in the files of my employer, the Ohio Division of Natural Areas and Preserves. Among other things contained in it, I found an initial list of natural arches in the state compiled by James L. Murphy in the mid-1970s. So, I was not the first one to be interested in studying Ohio's arches after all!

In part, the amount of time this survey has taken is a result of its being a personal project with work limited mainly to weekends and other spare moments. A more important — and exciting — reason is that the number of reported potential natural arches kept expanding faster than my ability to document them. Somewhere along the way natural pillars intruded into the study. The two features are often found in the same general area because the conditions required for their formation are similar. In researching arches, I often came across pillars, chimneys and tea tables, and it seemed natural to extend the study to them, especially since no comprehensive list of these structures appeared to have been made. Tracking down all of these leads gave me an excuse to explore some of the wildest and most scenic places in Ohio, and thereby greatly increase my already considerable appreciation of the state's natural wonders.

The work of recording Ohio's unusual geology is not yet finished and most likely never will be. New leads still appear now and then, coming from informants or dug out of obscure references hidden on library shelves. The appearance of this book will no doubt lead to the appearance of even more records, and they will be welcome. The most important product of this study is not a lengthy list of Ohio's arches and pillars, but the realization that these features play a larger role in our landscape than previously has been recognized.

Rainbows of Rock is a survey of the natural arches and pillars of Ohio — their numbers, types, possible means of formation and the stories of human interaction with them, both as recorded history and as handed down as legend. If this book increases public awareness of these valuable features and prompts a greater interest in learning about and protecting them, then the work will have served its purpose. Arches and pillars may be made of stone, but they are subject to human degradation and destruction nonetheless.

This book is not, however, a detailed guide to visiting every natural arch and pillar of record in Ohio. While a good number of these features are publicly owned and open for visitation, although restrictions may apply, many are found on private property. To protect their owners' rights and privacy, detailed location data are not given for these sites.

For those who are interested in visiting the publicly accessible arches and pillars, Chapter 7 — Publicly Accessible Natural Arches and Pillars of Ohio — provides helpful information that will make each visit easier and more meaningful than it otherwise might be. The publicly accessible features are marked with an asterisk (*) after their feature number in both the catalogs of arches and pillars and appendixes I and II. Take advantage of these opportunities to visit Ohio's arches and pillars and get to know some of the most delightfully scenic spots in the state. Be aware, however, that many of these features are in places where the footing might be unstable, where water or snow might have made the surface slippery, where talus might have accumulated and where steep slopes or vertical rock walls exist —

any or all of which can present risks of personal injury or death. So, when visiting these sites, please be careful, stay on marked trails or other routes intended for public traffic and follow other posted instructions. Lastly, please remember to treat these features gently. No one else is interested in knowing the date of your visit or who your love of the moment happens to be, so please leave all thoughts of graffiti, litter or campfires behind. Of course, if you are interested enough to hand over hard-earned cash for this book, you are most likely also concerned enough to leave the places you visit looking as good as or better than when you arrived. Thank you.

Some of the illustrations are first referred to in the text out of their proper order. The need to use some of the features for purposes of discussion before they are presented in the catalog where photographs of most of them are found led me to adopt this unconventional method.

A work such as this could never be accomplished without the help of many people. While I hesitate to offer a complete list of those who gave freely of their time and advice since many valuable contributions will almost certainly be accidentally overlooked, I would be remiss not to acknowledge those whose aid went beyond the call of duty. Robert Glotzhober of the Ohio Historical Society never lost interest in what I called the Ohio Natural Arch Survey and it was through his efforts that I finally made the contact that led to consideration of this work for publication. The day he spent showing me the arches of Fort Hill State Memorial led to several research permits graciously authorized by William T. Schultz and later by Bob himself which allowed me to study the Baker Fork Gorge section of the memorial in greater detail. Mike Hansen of the Ohio Geological Survey offered encouragement at a critical time and also reviewed the manuscript. Other reviewers included James L. Murphy who also shared information and pictures; Henry Wede, a fellow Midwestern natural arch aficionado; and Katryn Renard who made the mistake of offering to be my "ordinary reader," but turned out to be anything but ordinary in the interest and attention she gave to the work. Staff members of several agencies and organizations reviewed sections of the manuscript dealing with areas they manage to ensure that the information I presented was correct.

Ralph Ramey and Guy Denny, at various times my bosses at the Ohio Division of Natural Areas and Preserves, were always and continue to be interested accomplices. The division's Database Section not only gave me access to their files on natural bridges, but also kept asking me to add my accumulating information to it, impressing on me that the results of my efforts were important. Lisa Van Doren guided me through the files of the Ohio Division of Geological Survey and provided me with some very important illustrations. Horton Hobbs III of Wittenberg University in Springfield graciously opened up his Ohio Cave Survey files to me. Charlie Foster of the Ohio Division of Forestry provided information and a pleasant day of hiking to show me the arches in Hocking State Forest which he managed at the time. I'm sorry, Charlie, that I had to rename your arch.

Dale Liebenthal, Don Guy, Nate Fuller and others of the Ohio Geological Survey's Lake Erie Section introduced me to the Amherst arches and then allowed me to join one of their work cruises along Marblehead Peninsula and among the islands of western Lake Erie, patiently waiting while I took the photographs and measurements I needed. Herb McGuire and John Litten, successive managers of Camp Christopher, allowed me to visit the arch on their property and even helped me find it. Bob Stoll, biologist at the Ohio Division of Wildlife's Waterloo Experiment Station, not only shared information with me, but also cheerfully got my car out of a narrow ditch that opened up when I pulled off the road to park.

Martin McAllister, Jeff Johnson and Marilyn Ortt, all with the Ohio Division of Natural Areas and Preserves, freely provided both information and time to show me features they thought would be of interest. William Borovicka of the Vinton Furnace Experimental Forest led me to Arch Rock. Peter Whan gave me an informative tour of the Edge of Appalachia Preserve owned by The Nature Conservancy. Jeff Knoop opened doors and led me to Chapel Ridge Natural Bridge. Grant Thompson who, along with his siblings, owns the Amherst Arches site, gave me an enjoyable tour of the area. The many landowners who graciously allowed me to study the very special features under their care have my deepest gratitude.

Special thanks go to Mark Howes, fellow manager with the Ohio Division of Natural Areas and Preserves whose early comment sparked this whole project and who has continued to support it with new leads. And then there is Preston Fettrow. At an age when most men are content to rest on their laurels, Preston became engaged in a quest similar to mine — to photograph all of Ohio's natural arches. His generous sharing of photographs and information has been of immense help in compiling this report. Stewart Cearley, my faithful "cyber ghost," brought a recalcitrant computer to heel on several occasions and thereby saved my sanity. Jerry McDonald of McDonald & Woodward Publishing Company cheerfully guided me through the publishing process and saved me from numerous embarrassments. My mother, Marjorie E. Snyder, and sister, Judith A. Snyder, have been constant sources of encouragement for which I will be eternally grateful.

Introduction

The intent of my informal Ohio Natural Arch Survey was to locate, measure, map and photograph every natural arch located within the State of Ohio. The naivety of this goal was quickly exposed by the rapid addition of what were potential arches to the original short working list. While the goal of this informal Ohio Natural Arch Survey remains unfinished, *Rainbows of Rock* can be viewed as an interim report, one part of a work in progress. The sheer logistics of searching Ohio's 41,330 square miles precludes the chance of a thorough job being done by one person. Without a coordinated effort, work by multiple researchers would inevitably lead to more promising regions being searched several times while others of less interest would be overlooked entirely. In broken country, a single pass is not adequate. The Hocking Hills hollows, for instance, require at the least a trip along the rim of the cliffs and another along the base, and in many cases, where feasible, a third between the rim and base to ensure more nearly complete coverage, all of which greatly increases the distance that must be walked. Add to this the increasingly dense screen of vegetation cover that develops as formerly open land is allowed to revert to forest, especially in the more rugged parts of the state where natural arches and pillars are most likely to be found, and it becomes quite obvious that the search for Ohio's unusual geology will remain unfinished for a long time.

While a simple list of Ohio's natural arches and pillars with their dimensions and other raw data would no doubt be useful, it would not be all that interesting. Its major contribution would be to show that these unusual features do indeed play an important role in Ohio's landscape and merit further study. The basic framework of this report is just such a list, but it is intended to be much more. By including information on how these features may have formed and their place in the environment of which they are a part as well as the story of human interactions with them, it is hoped that this work will interest a broad spectrum of the public, making them aware of these natural treasures and encouraging them to work for their protection.

Previous Studies

Natural arches and pillars, especially the larger ones, have long attracted popular notice. The methodical study of them, however, is a fairly recent development. This may have been due in part to the perception of these features as uncommon "freaks of nature," insignificant features which offered little to advance the science of geology.

One of the first natural arches in North America to gain notoriety was Natural Bridge in Virginia. An early owner, Thomas Jefferson, published a description of it in 1787 which attributed the formation of the gorge which it spans to a cataclysm. The geologist Francis William Gilmer, who visited Natural Bridge with Jefferson, published his own version of its formation in 1818, becoming the first to recognize that it was formed by subterranean drainage in a cave system, much of which had subsequently collapsed.

As knowledge of the existence of Natural Bridge became widespread, similar features in other locations began to receive attention. In 1839, John C. Trautwine described a natural bridge in Tennessee and attributed its formation to cave collapse. In 1898, Arthur M. Miller discussed the formation of natural arches in Kentucky by headward erosion.

During the last few decades of the nineteenth century, exploration of the Colorado Plateau and surrounding areas in Utah, Arizona, Colorado and New Mexico revealed the large number of natural arches and pillars that occur there. The pace of reporting and theorizing about these fascinating features quickened. Public attention was captured by descriptions of the large natural bridges and arches of White Canyon and the Moab region in Utah, and spectacular Rainbow Bridge in

Arizona. As a result of this increasing interest, all three of these sites were given protection as national monuments or parks early in the twentieth century.

In 1910, Herdman F. Cleland published a seminal report in the *Bulletin of the Geological Society of America* entitled "North American Natural Bridges, with a Discussion of Their Origin." This was one of the first attempts to give the entire subject an organized explanation. Although Cleland does not claim to include all the natural bridges in North America, he does describe thirty-eight, including one in Indiana and "sandstone bridges" in Kentucky, but none in Ohio. Even so, he is still one of the first researchers to attempt to compile a master list for the entire continent. Before the advent of the internet, the most comprehensive published list of North American natural arches was the one compiled by Robert H. Vreeland in a series of guidebooks entitled *Nature's Bridges and Arches* published at intervals beginning in 1976. Most such efforts, however, were limited to a smaller area. Corgan and Parks's 1979 survey of the natural bridges of Tennessee and Stevens and McCarrick's 1988 study of the arches of Arches National Park are two examples.

The steadily growing interest among a small but increasing number of arch aficionados culminated in 1988 with the formation of The Natural Arch and Bridge Society (NABS), an organization having the expressed purpose of "increasing the understanding, appreciation and preservation of nature's most fascinating and beautiful landforms." This organization sponsors conventions and field trips, maintains a web site (*www.naturalarches.org*) and publishes a newsletter appropriately named SPAN. Its area of interest encompasses the globe and its growing list of natural arches is almost certainly the most comprehensive ever compiled.

Interest in Ohio's arches and pillars followed the national model. The first extensive notice taken of the Ohio features occurred in the 1889 edition of Henry Howe's *Historical Collections of Ohio* (Figure 1). Here they form part of a vast panorama of "history, geography, biography and travels." Their inclusion in Howe's work can be seen as part of a growing national awareness of the value of the natural world. There are, however, two other factors which played a part.

Figure 1. Rockbridge had become a tourist attraction by the middle of the nineteenth century. Note the photographer on top of the natural bridge in this old woodcut (Howe, 1896).

First, by 1880, most of the United States had passed well beyond the frontier stage. If much of the population was not wealthy, it was at least comfortable, and could indulge in occasional travel to interesting places. For most American citizens, the historically preferred destinations in Europe were still out of reach. There were, however, plenty of interesting things to see right here at home. The establishment of national "pleasuring grounds" such as Yellowstone National Park reflected a desire to both protect the country's outstanding natural features and to make them available to its people. This interest in nationally important scenery led inevitably to a rising awareness of scenic features on a local level. The development of railroads and interurban lines made state-wide travel easy and relatively inexpensive. Howe's inclusion of natural arches and pillars in his book gave it the dual function of travel guide and promoter of Ohio's place in the list of states having scenery worthy of attention.

The second factor involved in this rising interest in the natural arches and pillars of Ohio was their increased visibility. At the beginning of the Euro-American settlement of Ohio, old-growth forest covered some 25 million of the 26 million acres within the state's borders.

Elimination of the old-growth forest in favor of agriculture began almost immediately and continued until the middle of the twentieth century when only one million acres of large timber and two million acres of growth under 11 inches in diameter remained (Gordon, 1969). By 1900, the state had lost nearly all of its covering of primeval forest and its rocky foundations were widely exposed. Interesting features such as pillars and natural arches located within view of roads, canals and other transportation corridors were readily noticed. This visibility is emphasized in Van Tassel's *The Book of Ohio*, a large format collection of black-and-white photographs and essays published in 1901 to honor the state's centennial. In many of the pictures showing geological features, the absence of mature forest cover is striking. Since 1940, Ohio's forest has regained lost ground. This is especially true in the rugged Appalachian Plateau area of eastern Ohio where a number of the state's natural arches and pillars are found. As a result, many of them are far less visible now than they were a century ago.

While occasional articles about a natural bridge or pillar appeared in local newspapers or regional histories in the decades after Howe, the earliest attempt at a comprehensive list for Ohio did not occur until 1975 when James L. Murphy described several of the "about a dozen" he knew of in the state. At the same time, the Ohio Division of Geological Survey (Ohio DGS) and the Ohio Division of Natural Areas and Preserves (Ohio DNAP), both units of the Ohio Department of Natural Resources (Ohio DNR), separately began to compile lists of geologic features of interest. Information for these files came from previous publications, reports of field personnel and interested citizens.

One of the few discussions of natural arch formation relating specifically to Ohio was published by Thomas C. Kind in the 1976 proceedings of the Kentucky Academy of Science. In it, he describes Large (now called Ladd), Little and Independence (now called Irish Run) natural bridges and attributes their formation to the enlargement of vertical crevices and horizontal bedding planes at the face of waterfalls.

Also in 1976, Robert H. Vreeland published *Nature's Bridges and Arches, Volume 14: Midwestern States* as part of his national survey.

In it, he describes five arches in Ohio: Ladd (called Rock Bridge by Vreeland), Rockbridge, Needles Eye in Lake Erie, The Keyhole (an unnamed bridge in Vreeland) and Natural Y Arch (also unnamed). Three more arches (Beaver Creek Natural Bridge, Mineral Arch and presumably Mustapha Natural Bridge, none of them named) are listed as "Others" at the end of the book, giving a total of eight for the state. Two volumes of "Additions" added several more to the list for Ohio. Vreeland's entire list of named arches in the United States is summarized in the 1994 revision of his *Volume 1: General Information*. In it, Cats Den Bridge, Hagley Hollow Arch, Irish Run Natural Bridge and Rock House are included, bringing Vreeland's total for the state of Ohio to twelve.

In 1983, field work and literature searches for the present survey of natural arches and pillars in Ohio were initiated. Preliminary findings were reflected in an article entitled "Natural Bridges in Ohio" by Michael C. Hansen, published in the Ohio DGS newsletter in 1988 which listed forty natural arches. An article by the present author in the newsletter of Ohio DNAP during that same year mentioned several of these features which were open to public visitation and helped to raise public awareness of them, as did an illustrated page on the division's web site. A well-illustrated revision of Hansen's 1988 article appeared in the November–December 1993 edition of the Ohio Historical Society's magazine *Timeline*. By then, the number of known natural arches in the state had risen to "over forty." Dual articles by the present author and Henry Wede in *SPAN: Newsletter of the Natural Arch and Bridge Society* (2007) raised the number to seventy-four and brought Ohio's natural arches to the attention of the larger arch community. The present publication brings the total of natural arches recorded in Ohio to eighty-six.

While the literature on Ohio's natural arches may be meager, that on the natural pillars of the state is virtually non-existent. Howe (1896) and Van Tassel (1901) describe and illustrate several, most notably the Morgan County Devils Tea Table and Pompeys Pillar in Greene County. Joseph T. Harrison's article entitled "The Pillars of Harrison County" in the journal of the Ohio Archaeological and Historical Society (now the Ohio Historical Society) (1922) is apparently the earliest attempt at listing such features, although it covers only one county of the state. The peripatetic James L. Murphy listed and illustrated several natural pillars, which he called standing stones, in eastern Ohio and discussed their archaeological potential in an article published in *The Ohio Archaeologist* in 2004. Information on others has been provided by articles in local newspapers and histories, by local informants and by several government and private natural resource agencies. This publication records eighteen natural pillars in Ohio and represents the first attempt at a comprehensive listing and discussion of these features.

Methodology

The initial phase of this study involved the compilation of a list of potential natural arches found within Ohio. Early additions to this list came from Henry Howe's *Historical Collections of Ohio* and Van Tassel's *The Book of Ohio* which contain pictures and short descriptions of interesting geological features. Further additions came from newspaper and magazine articles, files in the offices of Ohio DNAP and Ohio DGS and from conversations with field personnel in several state and private agencies. Pictures submitted to the Geological Survey's photo contest led to several interesting additions. Eventually, a second list, this one containing potential pillars, was compiled.

Once it was determined that there was a number of these interesting features to be found, an attempt was made to visit each one. Some information sources supplied location data; others, most notably the photo submissions to the Geological Survey contest, had little or no information beyond a name. Success in tracking down these more problematical features was often a matter of fortuitously showing the picture to the right person — sheer good luck.

When the location of a feature was determined, the site was visited if possible. The feature, if found, was photographed from several angles on both slide and black-and-white film, and later, digitally. A foot-long ruler with the inches marked alternately black and white was often included to provide scale. The location of the feature was noted on a copy of the relevant topographic map; later in the survey, GPS coordinates were also acquired.

Measurements of the feature were made with a 100-foot-long cloth-wrapped nylon line with each foot marked in color-coded permanent paint. The end was secured in the ground by an aluminum tent stake, allowing the line to be pulled taut. An old plumb bob tied to the same end allowed the line to be dropped to get vertical measurements. While stretching of the line may have introduced some variation in these measurements, especially in the longer vertical ones, such variation was small in relation to the lengths measured. Late in the survey, a 100-foot-long reeled cloth tape marked in feet and inches was used. Lengths of less than 12 feet were sometimes measured with a 6-foot folding wooden rule or a 12-foot metal tape. Where a vertical measurement had to be made from the ground up, the metal tape or folding wooden rule were used; in some cases it was necessary to add the known 6-foot-2-inch height of the measurer to arrive at the final figure. While the measuring devices used may not be as accurate or convenient as more advanced means which have recently become available, the figures arrived at by their use are felt to be adequate.

The English measuring system of feet and inches was used to coordinate the results with those found in other studies of natural arches consulted at the beginning of the project. The ready availability of measuring equipment utilizing this system and the author's familiarity with it were also factors. Measurements in this report are given in feet so as to honestly and accurately reflect those made in the field and to make them more easily comprehended by the majority of potential readers. An English-metric conversion table is provided in Appendix III for those desiring metric equivalents.

In spite of the care taken with these measurements, a second researcher or even the same one returning at a later day could easily get slightly different results. There are a number of possible reasons for this variation, not the least of which is the difficulty of determining exactly what to measure. This is due mainly to the irregularity of the features being measured. Locating the end of a tape measure a few inches in either direction can make a large difference in the final measurement. An attempt was made to measure every listed arch and pillar myself in order to limit confusion and ensure consistency.

The measurements along with any other pertinent information were recorded on rough field sketches made on-site during the visit. These sketches usually included a ground plan and cross-sections of the feature. Depending on its complexity, drawings could also be made to clarify the photographs. Many of the sketch maps were later converted to scaled drawings on graph paper. With the location of the feature determined and the information collected on-site in hand, it was then possible to research its geology and history in more detail. The information gathered was then compiled into a summary, the combined total of which forms the Catalog of Ohio's Natural Arches and the Catalog of Ohio's Natural Pillars presented in chapters 3 and 5, respectively.

All features in the catalogs were numbered. Several previously existing cataloging systems, including one utilizing GPS data, were considered, but the one finally adopted recommended itself by its simplicity. A feature number consists of four parts separated by hyphens; the catalog number for Riverbend Arch in Adams County, for example, is OH-A-ADA-01. The four parts of the catalog number are, first, OH for Ohio, followed by a capital A for arch or P for pillar, then the three-letter Ohio Department of Transportation abbreviation for the county (Figure 2; Table 1) in which the feature is found and finally a two-digit number. Arches and pillars were numbered in sequence within each county in no particular order. The great deficiency of this system is that a current copy of the list is needed to ensure that numbers are not duplicated when newly found features are added to it. The master copy of the catalog is filed with Ohio DNAP.

All features in the catalogs were also named. While numbers might suffice, they do not create a clear picture of the feature being discussed and so can be a source of confusion. They are also boring. Names are much easier to grasp and have been the preferred method of designating features of interest for millennia. They can also provide immediate information such as general form or type. In this report, previously published names were given preference unless they failed to conform to the criteria given below. In cases where more than one published name existed for a given arch, the one which most closely matched the naming criteria was used. When no published

Figure 2. A map showing the locations of Ohio's counties. See Table 1 for the names of counties corresponding to the numbers shown on this map.

name for a feature could be found, but a local name was in use, this was utilized. Whenever the designation "natural bridge" or "arch" was incorrectly used in a previously existing name, the designation was corrected, but the name itself was retained. For example, the name of the feature in Fort Hill State Memorial originally called Y Natural Bridge was changed to Y Natural Arch because the feature does not fit the accepted definition of a natural bridge. If no name was known, then an original name was provided. Arches were often named after the nearest watercourse, valley or geological feature. This procedure for providing a name follows a tradition well established in the scant literature on Ohio's natural arches. Sometimes the appearance of the

Table 1. Names of Ohio Counties

1 Adams		45 Licking
2 Allen		46 Logan
3 Ashland		47 Lorain
4 Ashtabula		48 Lucas
5 Athens		49 Madison
6 Auglaize		50 Mahoning
7 Belmont		51 Marion
8 Brown		52 Medina
9 Butler		53 Meigs
10 Carroll		54 Mercer
11 Champaign		55 Miami
12 Clark		56 Monroe
13 Clermont		57 Montgomery
14 Clinton		58 Morgan
15 Columbiana		59 Morrow
16 Coshocton		60 Muskingum
17 Crawford		61 Noble
18 Cuyahoga		62 Ottawa
19 Darke		63 Paulding
20 Defiance		64 Perry
21 Delaware		65 Pickaway
22 Erie		66 Pike
23 Fairfield		67 Portage
24 Fayette		68 Preble
25 Franklin		69 Putnam
26 Fulton		70 Richland
27 Gallia		71 Ross
28 Geauga		72 Sandusky
29 Greene		73 Scioto
30 Guernsey		74 Seneca
31 Hamilton		75 Shelby
32 Hancock		76 Stark
33 Hardin		77 Summit
34 Harrison		78 Trumbull
35 Henry		79 Tuscarawas
36 Highland		80 Union
37 Hocking		81 Van Wert
38 Holmes		82 Vinton
39 Huron		83 Warren
40 Jackson		84 Washington
41 Jefferson		85 Wayne
42 Knox		86 Williams
43 Lake		87 Wood
44 Lawrence		88 Wyandot

feature resulted in its name, as in Castlegate Arch. A few names such as Rockgrin Arch are more whimsical, but even serious researchers need to have a little fun now and then. No feature was named for a living person. Fosters Arch, the one exception which was made early in the course of the survey, has been corrected and is now called Balcony Natural Bridge. The only feature named for a person no longer living is Tecumseh Arch, although several features in the Hocking Hills named for the hollows in which they are found do carry the local family names applied to those valleys. Governmental unit names such as those for townships and counties were avoided when possible since the chance of finding more than one feature in such a unit creates the potential for confusion. A few such names that were already firmly established, such as Pike Arch named for Pike County, were retained, even after the discovery of more arches in the county.

The catalogs have been arranged by geological periods which permits the discussion of each feature in relation to the rocks of which it is formed and to other features formed in the same strata. At the same time, this arrangement provides a general overview of the geological history of Ohio. The two measurements considered most important for a feature are given at the head of each entry in the catalog in both English and metric units to facilitate comparison with other entries. For arches, these are the span and the clearance which together give an impression of the size of the opening and therefore of the arch's defiance of gravity. For pillars, the measures of height and width are given.

While most of the larger arches and many of the smaller ones are included here, there are certainly others hidden away in deep hollows which have not been documented. For pillars, the number of potential additions is even greater. The two lists presented here can therefore only be considered preliminary. It is suspected that their publication will encourage those who are aware of similar features to report them. Should you know of a natural arch or pillar which meets the criteria for listing, please notify the Ohio Division of Natural Areas and Preserves, 2045 Morse Road F-1, Columbus, OH 43229-6693. Please include the feature's estimated size, specific location, ownership if known and a photograph if available.

Ohio's Natural Arches and Pillars
in Geographic Perspective

One of the more important conclusions resulting from this study is that natural arches and pillars occur more frequently in Ohio than was previously thought. While not common, they are abundant enough to be considered more than geological freaks. They are natural results of the interaction of conditions and processes which come together in a number of locations across the state. Where these conditions are most favorable, arches and pillars can be found in clusters. Where conditions are less favorable, these structures occur as isolated features or not at all.

How does Ohio's collection of arches and pillars compare with those found in the rest of North America? Putting the results of this study into a regional context shows that Ohio's arches and pillars are part of a larger collection of such features which occupies the greater part of the continental United States. Just as favorable combinations of conditions result in local concentrations of arches and pillars within Ohio, so favorable conditions on a continental scale result in regional concentrations.

Two of the most notable of these concentrations are the clusters of arches and pillars found in and near Arches National Park in Utah and in the Red River Gorge region of Kentucky. The arches of these two areas have received more study and publicity than those in any other region. As a result, similar geological features found elsewhere are thought by many to be a rare occurrence and something of an anomaly. In fact, arches and pillars are widespread phenomena whose apparent rarity is more a consequence of the lack of attention given them than a reflection of reality. When they are sought, they are found.

No one has tried harder to give natural arches their due recognition on a continental level than Robert H. Vreeland. His series of regional guides to the arches and natural bridges of the United States includes twenty-three slim volumes covering every part of the nation. If nothing else, his work proves that natural arches are not limited to two major sites, but are found in almost every state. However, even Vreeland's monumental work does not give a complete picture of the number and location of natural arches. He lists only twelve for Ohio whereas the present report includes eighty-six. If this ratio holds true for other areas, the number of arches must be large indeed. By way of comparison, Vreeland (1976) lists seven arches in Michigan, two in Indiana, four in Illinois and one in Pennsylvania. Corgan and Parks (1979) discuss thirty-six natural arches found in Tennessee.

Insofar as pillars are concerned, little work other than the listing of these features has been done in most areas. Geyer and Bolles (1979, 1987) describe eleven features in Pennsylvania which might be considered natural pillars. At least four named pillars are known in Indiana. Except for unusually large or picturesque examples, natural pillars have not been inventoried or studied to the extent they deserve. If so many pillars have been found in Ohio, a state which at best can be considered "partly hilly," then other states having more dramatic topography probably contain numerous such features awaiting recognition and study.

Chapter 1

Conditions Required for the Formation of Natural Arches and Pillars

To our eyes the landscape appears eternal. In reality it is as fluid as the water flowing over it. The only difference is the time scale in which the changes occur. Were we able to speed up geologic time and watch the landscape as though it were captured in time-lapse photography, we would find it rolling and heaving as the forces of erosion — rain, humidity, glaciers and the sea, wind and the stress of extreme temperature changes — wear away what tectonic processes raise. Plants, animals and humans play a part — humans often a very obvious part — in eroding the bedrock foundation of the landscape.

Although the specific actions of these forces are as varied as the situations in which they work, their contributions to erosion can all be placed into one of two broad categories: mechanical weathering or chemical weathering. Mechanical weathering involves breaking the rock into ever smaller pieces by the application of physical force and can be accomplished by, among other things, the pressure of expanding water as it freezes in cracks, abrasion by wind- or water-carried particles or by impact at the end of a gravity-induced fall. Chemical weathering results in a change in the rock's chemical makeup and can lead to its disintegration, such as by dissolving some of the minerals which then wash out of the rock or by forming entirely new minerals which may have a greater volume than the original material, thereby creating stress within the rock. When mechanical and chemical weathering occur as a result of exposure of the rock to the action of the atmosphere, the collective process is called atmospheric weathering or atmospheric erosion.

Chemical weathering is far more active under the conditions found on Earth than is mechanical weathering, although its effects are more subtle and easily overlooked. Neither kind of weathering operates uniformly. Differences in rock composition, variations in slope and climate and the amount of time the agents of erosion have had to work on a given piece of bedrock all help to determine how much and in what ways it will be eroded. On a large scale these variations result in hills and valleys, cliffs and caves, basins and plains. On a smaller scale, they are responsible for slump blocks, waterfalls, sink-holes, nooks and crannies.

Natural arches and pillars are part of the collection of small-scale features found in the landscape and they represent a surprisingly common result of the interplay between the forces of erosion and the bedrock upon which those forces work. Arches and pillars are not found everywhere, however. In order for them to form, certain factors must be present — consolidated rock, an erosive agency, to-pographic relief, openings in the rock and time.

Consolidated Rock

If an arch or a pillar is going to stand, it must be made of rock that is solid enough to hold together against gravity. Under certain conditions, all rock can form arches or pillars. Those formed in weak rock such as shale, however, are small and short-lived (Figure 3). In Ohio, all arches and pillars of significant size have formed in sand-stone, conglomerate, limestone or dolomite. If a given rock is going to form such a feature, it must be strong enough to hold up its own weight. This is determined in part by the thickness of the beds composing it. Thin layers of rock will tend to form thin arches which are weaker and less resistant to erosion than are thicker ones. Arches formed in such rock will also have a shorter span than those formed in thicker layers.

All bedrock exposed at the surface in Ohio is sedimentary in origin; that is, it formed through the accumulation and consolidation

Figure 3. An example of a natural pillar formed in shale (Van Tassel, 1901). The thinness of the layers making up the rock along with the weakness of the shale itself ensures that this feature will not survive long, especially given its location on the wave-assaulted shore of Lake Erie.

of sediment, usually, but not always, in water. Sandstone, limestone and shale are common sedimentary rocks in Ohio. Rock formed by solidification from a molten state, such as granite and basalt, is classified as igneous. Metamorphic rocks such as gneiss, quartzite and marble have been changed from their original state by the extreme temperature, pressure and chemical conditions found deep below Earth's surface. Igneous and metamorphic bedrock is found in Ohio, but it is buried under hundreds of feet of sedimentary rock that has been deposited on top of it. Even so, it is still possible to find examples of igneous and metamorphic rocks scattered over three-quarters of the surface of the state, carried southward from Canada by continental glaciers during the Pleistocene.

Sedimentary rock made of fragments broken or worn from previously existing rock is called "clastic." Sandstone, conglomerate and shale, differentiated by the size of the fragments of which they are formed, are examples of clastic sedimentary rocks. Calcareous sedimentary rocks such as limestone and dolomite form from the chemical precipitation of calcium carbonate in oceans or large lakes, or by the accumulation of shells and other calcareous remains of animals. Sedimentary rock can also form from the deposition of plant debris in swamps (coal) and through chemical action or evaporation (rock salt, gypsum).

Sedimentary rocks are characteristically layered. These layers are assumed to have been flat or nearly so when originally laid down, but may have been tilted, folded, raised, broken or even overturned by various earth movements over time. This layering of sedimentary rocks is the principle tool used by geologists to determine their relative ages. Since each layer is presumed to have originally been deposited as a generally flat sheet, it follows that a given layer of rock must have been deposited after the layer beneath it was in place and before the layer above it was laid down. It must then be younger than the lower layer and older than the upper layer.

This sequence of layers is interrupted at intervals by major breaks called "unconformities" created during times when conditions were not conducive to the deposition of sediments. Most unconformities mark a time of erosion in which previously deposited rock was removed. The more drastic of these interruptions can be followed across continents and have provided the means of systematically dividing the stacked series of sedimentary rocks into units. The fossil remains of plants and animals which often vary in both type and abundance from one layer of rock to another were found to be equally useful in determining which unit a given bed of rock belonged to. Ordering the rocks led to dividing geologic time itself into eras (Figure 4). From lowest (oldest) to highest (youngest), these eras are Precambrian, Paleozoic, Mesozoic and Cenozoic. The last three eras are subdivided into periods. As seen in Figure 4, not all of these eras are represented at the surface in Ohio. The igneous and metamorphic Precambrian foundations of the state as well as the ancient sedimentary rocks of the Cambrian Period are covered by younger rocks. Strata from the Mesozoic Era and part of the Cenozoic are missing entirely, either because they were never deposited or because they have eroded completely away.

Years before present, in millions of years	Eras and duration, in years	Periods and duration, in years	Area of outcrop in Ohio and principal rock types
1.6	CENOZOIC 66+ million	QUATERNARY 1.5-2 million	northwestern 2/3 of Ohio– unconsolidated sand, gravel, clay
66.4		TERTIARY 62.5 million	NOT PRESENT IN OHIO
144	MESOZOIC 179 million	CRETACEOUS 78 million	
208		JURASSIC 64 million	
245		TRIASSIC 37 million	
286	PALEOZOIC 325 million	PERMIAN 41 million	southeastern Ohio–shale, sandstone, coal, clay, limestone
320		PENNSYLVANIAN 34 million	eastern Ohio–shale, sandstone, coal, clay, limestone
360		MISSISSIPPIAN 40 million	east-central, northeastern and northwestern-most Ohio–shale, sandstone, limestone
408		DEVONIAN 48 million	central, northeastern and northwestern Ohio–shale, limestone
438		SILURIAN 30 million	western Ohio–dolomite, limestone, shale
505		ORDOVICIAN 67 million	southwestern Ohio–shale, limestone
570		CAMBRIAN 65 million	NOT EXPOSED IN OHIO — Cambrian sandstones, shales, and carbonates and Precambrian sedimentary, igneous, and metamorphic rocks present in subsurface
		PRECAMBRIAN 3,400 million	

Figure 4. The units of geologic time that are represented in the rocks of Ohio. Wavy lines indicate major unconformities. (Use courtesy of Ohio Department of Natural Resources, Division of Geological Survey)

Nearly all of the exposed bedrock in Ohio was laid down during the Paleozoic Era. Our share of the geologic layer cake includes rock laid down in the Ordovician, Silurian, Devonian, Mississippian, Pennsylvanian and Permian periods. After that time, Ohio was high enough that erosion, rather than deposition, was the dominant process shaping the surface of the region. Not until the last part of the Cenozoic Era during the Pleistocene Epoch of the Quaternary Period, commonly referred to as the Ice Age, were sedimentary deposits laid down once again.

19

The fact that rocks of six different periods are exposed at the surface in Ohio can be attributed to the presence of the Cincinnati Arch, one of the major geologic structures of the Midwest. Although called an arch, the Cincinnati Arch is not a natural arch of the type documented here, but an elongated bedrock dome whose form can be traced for 600 miles from Alabama to Ontario (Figure 5). Its central

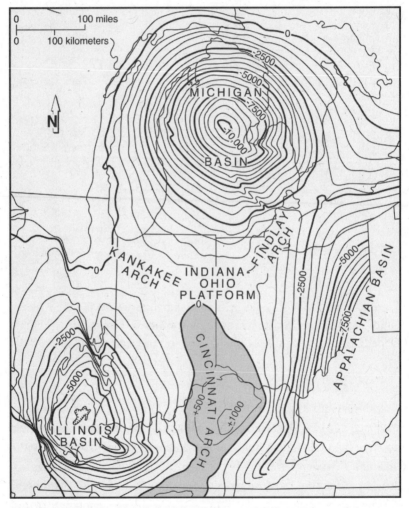

Figure 5. The Cincinnati Arch as it appears at the surface of the now-buried Trenton Limestone. The contour interval is 500 feet. (Use courtesy of Ohio Department of Natural Resources, Division of Geological Survey)

axis runs north-by-northeast, entering Ohio from the south near Cincinnati. In west-central Ohio the axis splits into the Kankakee Arch, which angles northwest across Indiana, and the Findlay Arch, which continues the northerly trend of the main axis through Marblehead Peninsula and the Erie Islands into Ontario where it drops below ground level.

In form, the Cincinnati Arch resembles an anticline in which originally horizontal strata have been bowed upward. More likely, the crest of the arch represents a stable area which resisted the downwarping responsible for forming the Appalachian Basin to the east, the Illinois Basin to the west and the Michigan Basin to the north. These basins have been completely filled with younger sedimentary rock strata. This filling coupled with long periods of erosion has leveled any topographic relief the arch might have originally possessed at Earth's surface. Its form is still present deep underground, however. Figure 5 shows what the surface of the Midwest would look like if all overlying sediments down to the level of the Ordovician-age Trenton Limestone were stripped away. The three basins defining the Cincinnati Arch and its two branches are readily visible. At the surface, the Cincinnati Arch is revealed by the successively younger layers of rock filling the basins. As represented on a map of the surface geology of the region, it has the classic appearance of a breached anticline — an elongated horizontal bull's-eye with the oldest rocks in the center surrounded by concentric rings of increasingly younger rock outward from the center (Figure 6).

Ohio occupies part of the northeastern quadrant of this bull's-eye (Figure 7). The oldest rocks found at the surface within the state appear around Cincinnati near the centerline of the arch. These are the richly fossiliferous Ordovician limestones and shales for which the region is famous. Surrounding this central area is a band of younger Silurian rock, mostly dolomites, which underlie the western Ohio plain. Next come rocks of the Devonian, Mississippian and Pennsylvanian periods which form bold north-south stripes across the state from west to east. The youngest bedrock found in Ohio is Permian in age and occupies a small area along the Ohio River in the southeastern part of the state. Since most of the Ordovician and Silurian rock deposited in

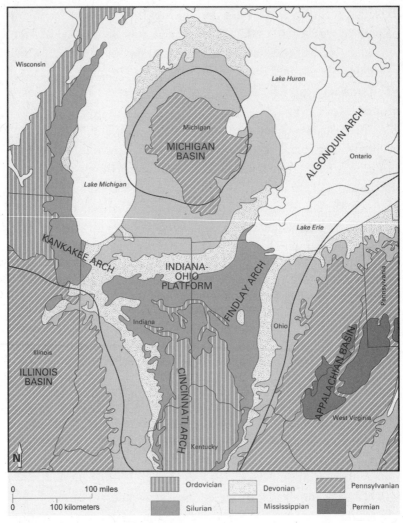

Figure 6. Surface geology of the Cincinnati Arch. (Modified legend; use courtesy of Ohio Department of Natural Resources, Division of Geological Survey)

Ohio is calcareous while much of the rock found in younger strata is clastic, the state is neatly divided into a western half dominated by limestone and dolomite, and an eastern half covered mainly by sandstone, conglomerate and shale. This division has a definite effect on the scenery of the state and on the natural arches and pillars which

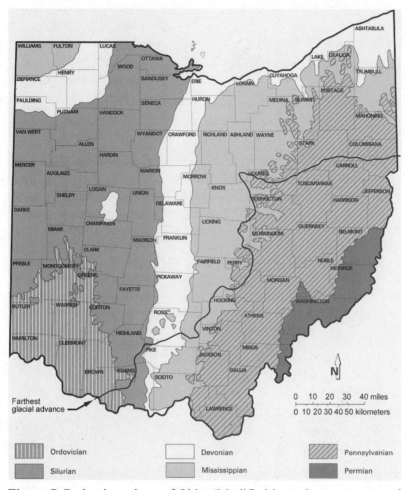

Figure 7. Bedrock geology of Ohio. (Modified legend; use courtesy of Ohio Department of Natural Resources, Division of Geological Survey)

form a small, but interesting, part of that scenery. Figure 8 gives an overview of the locations of Ohio's natural arches and pillars in relation to its surface geology.

The characteristics of each bed of sedimentary rock are determined by the inherent traits of the sediment which formed it and the conditions under which that sediment was deposited. The color, strength, resistance to erosion and even composition of an individual bed can change over distance both horizontally and vertically. A bed

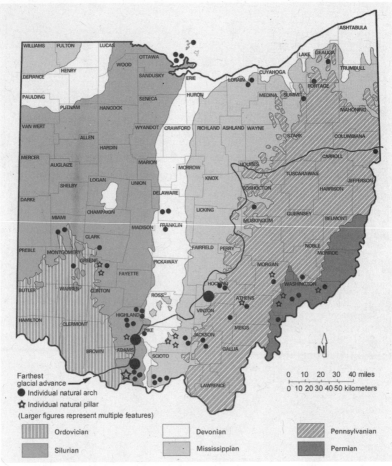

Figure 8. General location of Ohio's natural arches and pillars in relation to bedrock geology and the glacial boundary. Large circles designate multiple arches (Highland County: 8 arches; northern Adams County: 12 arches; western Hocking County: 18 arches). The large star in Adams County represents 4 pillars.

of conglomerate, for instance, may be found grading into sandstone and then into shale as it is traced cross-country. Sometimes rock layers pinch out altogether and are replaced by different rocks above or below which thicken to fill in the gap. All this variety creates the possibility of almost infinite variations in form as the exposed rock is attacked by the erosive agents of water, wind and temperature. It also means that a rock unit capable of forming arches and pillars in one part of the state may lose that ability within a few miles or disappear altogether. This accounts for the clustering of arches and pillars in certain areas. The Mississippian Black Hand Sandstone, for example, is strong enough to form arches. However, the deposit is found only in central Ohio between the Hocking Hills and Newark, and in the vicinity of Mohican State Park. Arches and pillars formed within it are therefore limited to these regions.

Erosive Agency

In order for rocks to be carved into arches and pillars, an agent of erosion must be present to do the carving. The most commonly mentioned erosive agents are wind, temperature change and water.

Wind has often been cited as a major factor in eroding rock, especially in arid lands where soil and vegetation are sparse, but wind action is limited to the height to which it is able to raise sand and other grit and throw it against rock. In most cases this will not exceed 2 to 3 feet. Even under optimal conditions, wind abrasion generally produces only small-scale features and these are usually found in rocks already weakened by chemical weathering from groundwater seepage. In the humid east where plant cover and ground moisture combine to hold sand in place, wind abrasion is a minor erosive force.

Temperature change, both daily and through the year, has also been listed as an agent of rock fracture. By itself, it appears to be most effective in reducing rock to fragments by heating and cooling the rock's surface, causing it to expand and contract and eventually break under the stress. The greatest contribution temperature change makes to erosion is its effect on water which it evaporates, precipitates, thaws and freezes. The importance of the effect of temperature on water should not be underestimated, for water is by far the most important agent of erosion. Much of its potency comes from its ability to carry substances in solution which increases its acidity and thus its ability to chemically alter any rock it contacts. Precipitation, which initially provides the water in most erosive situations, can vary chemically from neutral (neither acidic nor alkaline) to the acidity level of a tomato (pH 4.5). Much of this acidity is due to the presence of dissolved carbon dioxide. Rainwater can also carry nitrogen and oxygen as well as sulfuric, nitric and hydrochloric acids, all of which increase its weathering capabilities.

Once on the ground, water can increase its acidity still further as it passes through soil where plant respiration and decomposition make additional carbon dioxide available. Water carrying acids derived from the soil is capable of dissolving certain minerals, such as limonite, which are ordinarily insoluble. The contribution of soil acidity to rock weathering is especially great in humid regions such as Ohio where much of the water reaching bedrock has percolated through an often thick layer of decomposing plant matter and soil.

Armed with this increased acidity, water attacks rock through the processes of solution, oxidation, hydration and the formation of carbonates. In the process of solution, affected parts of the rock are dissolved into the passing water and carried away. Oxidation involves the addition of oxygen to a substance to form a new one; the rusting of iron is a common example. In hydration, water is added to the original substance, as in the change of anhydrite to gypsum. An example of the formation of carbonates is the changing of calcium carbonate found in calcareous rock to bicarbonate by water carrying carbon dioxide. In many cases, the new products formed by the last three processes are more soluble than the original ones and their removal compromises the integrity of the rock in which they are found. They may also occupy more space than the original materials and, by expanding, put stress on surrounding unaltered rock.

The extent and importance of this water-induced chemical weathering cannot be overstated. Most reactions in sedimentary rocks such as those found at the surface in Ohio take place in a wet environment which may be nothing more than a thin sheet of dampness coating the walls of a narrow fissure. Calcareous rocks are especially susceptible to such attack, a fact responsible for caves and other karst features. Siliceous sand grains are much less soluble, but the iron and calcareous cements holding them together to form sandstone and conglomerate are not so resistant, making these rocks also vulnerable to chemical weathering. Even in places where liquid water cannot flow, water is still able to perform its weathering action, reaching these sites as water vapor which condenses on the rock.

Water is also responsible for much of the mechanical weathering which occurs. As water freezes, it experiences a nine-percent increase in volume. Under natural conditions, this enables it to exert a force of up to 2,000 pounds per square inch, strong enough to crack granite and marble. For this to occur, the water must enter a crevice in the rock while in the liquid phase and then freeze. The crevice must be tortuous enough to prevent the expanding ice from pushing out through its opening and so dissipating its power. Such frost wedging is most effective where freeze-thaw cycles occur repeatedly. The process is important in Ohio which lies in an area of moderate frost-wedging potential.

At its most massive, ice in the form of glaciers is capable of mechanically eroding rock to an extensive degree, widening valleys and carving away mountainsides. During the Pleistocene, glaciers advancing into Ohio blocked the courses of existing streams, causing them to flood their valleys and form glacial lakes, some of which were of great size. When the rising water reached a low point in the surrounding valley wall, it poured over in a massive flood, carving a trench into the underlying bedrock in the process. This glacial diversion was responsible for creating most of the picturesque gorges cutting across Ohio's present landscape and so was directly responsible for creating the topographical conditions under which many of our natural arches and pillars were formed.

Liquid water also accomplishes mechanical erosion, mostly in the form of flowing streams carrying sand and gravel which abrade the bedrock confining it. In places of turbulent flow, such as at the base of waterfalls, bedrock can be broken and pulled out of place causing undercutting and eventual collapse of overlying beds. This is especially effective where the lower beds are weak and easily eroded. On a more subtle level, alternate wetting and drying of rock can also cause mechanical weathering through detaching flakes or causing cracks.

Finally, water is the agent most responsible for removing the products of weathering. It is water in the form of rivers and glaciers which carries most of the earth to the sea, whether mechanically as sand and gravel, or chemically as solutes dissolved within it. Wind can also play a part in transporting rock debris, but its contribution is limited to removing the finer particles and is most effective in arid climates where soil and plant cover do not protect the land surface. Gravity aids in transportation by causing loosened particles to move downward so long as the surface across which they are moving is too steep to stop that movement. Gravity also provides the energy for the movement of water, a contribution easily overlooked. In the final analysis, however, water is the greatest agent of transportation of eroding sediments and solutes.

Ohio receives between 30 and 40 inches of precipitation each year, most of it coming as rain during the summer months. This amount of water leads to fairly rapid erosion of rock and is one reason why some of our geological features tend to be smaller than those found in states farther west where precipitation levels are lower. Our arches and pillars are weathering away almost as fast as they form.

Topographic Relief

Relief can be defined as the variation in height of Earth's surface. While gravity gives water its erosive force, relief influences the effectiveness of that force by providing diverse surface gradients across which the water can flow. Relief not only affects water's ability to remove erosion debris, but also directly influences its erosive capacity. In areas of high relief where there are large vertical differences between the highest and lowest points on the surface, water will have the opportunity to fall a greater distance, giving it more energy and a greater erosive force which will be reflected in mechanical scouring, cavitation and similar processes. Conversely, areas of low relief where there is little vertical distance between the highest and lowest points on the surface will allow the water flowing over it little erosive force. Arches and pillars are always found in areas of relatively high relief, even though that relief may be expressed in an extremely limited area.

The amount of relief also determines the size of an arch or pillar. Obviously such a feature cannot be any higher than the immediate topographic rise in which it forms. The total amount of relief available within Ohio is 1094 feet. This relief is not all found in one place, however. It is spread over 100 miles between Campbell Hill in Logan County, the highest point in the state at 1549 feet above sea level, and the southwestern corner of Hamilton County where the Ohio River crosses our western boundary which, at 455 feet above sea level, is the lowest point in the state (Figure 9). In any one place, the amount of relief is much less than this potential maximum. The greatest relief expressed in Ohio is 700 to 800 feet found in scattered localities in Adams, Scioto and Pike counties where resistant Mississippian rock caps the hills and along the Ohio River in Monroe and Belmont counties where a preglacial divide created an elevated upland into which the Ohio River was trenched. In both areas, glacial diversion of drainage has caused streams to become deeply incised, creating areas with relatively high local relief (Figure 10).

Topographic features large enough to extend across counties or states make up regional relief as opposed to local relief which describes small-scale individual features such as the cliff in which an arch is found. The most important large-scale topographic feature found in Ohio is the continental divide that separates the Great Lakes-Saint Lawrence River watershed draining into the Atlantic Ocean from the Ohio-Mississippi River watershed draining into the Gulf of Mexico. This divide is a geologically recent feature, having taken its present form only after the retreat of the last continental glacier from the region and is formed, in part, of thick glacial deposits. In places it may follow the line of an earlier divide, but it crosses some preglacial and interglacial divides at right angles. This divide determines whether a stream will flow out of Ohio to the north or to the south and so has had an important, if distant, effect on the formation of arches and pillars. Since all the major streams in Ohio head on or near the divide, it also determines the greatest height from which they can fall. The other half of that equation — the lowest point or base level to which the streams can fall — is provided by the Ohio River on the south and Lake Erie on the north.

Several other regional topographic structures also have had an effect on the formation of arches and pillars in Ohio. Both the Allegheny Escarpment (called the Portage Escarpment where it parallels the Lake Erie shore) which is a noticeable rise running from the Ohio

River in Adams County to the Pennsylvania border in Ashtabula County, and the Niagara Escarpment, which curves around the south-western corner of the state, provide elevation changes of tens to hundreds of feet. Both mark the edges of resistant rock strata which now stand higher than neighboring areas that are underlain by weaker rock.

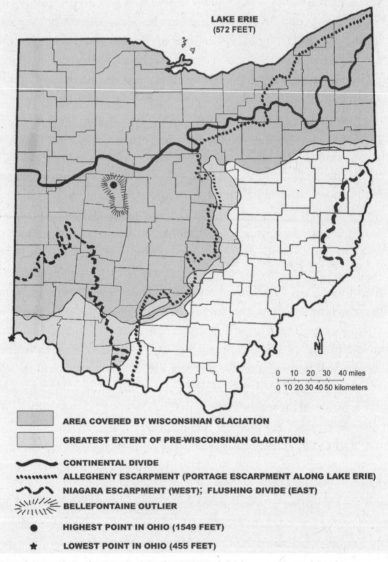

Figure 9. Major topographic features of Ohio mentioned in the text.

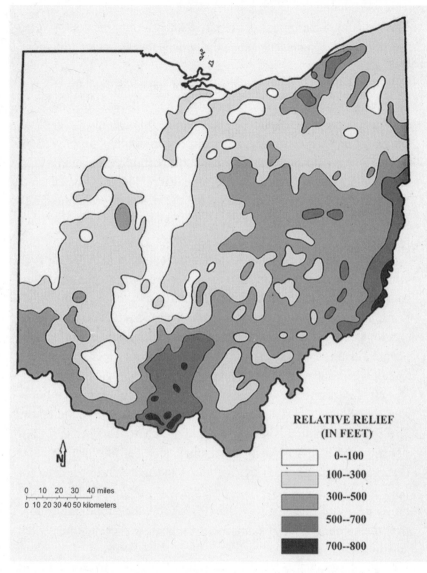

Figure 10. Relative relief of Ohio. The rugged southeastern half of the state contrasts sharply with the flat northwest. The largest area of high relief near the Ohio River marks both the edge of the Allegheny Escarpment and a region of extensive down-cutting by glacially diverted streams (see Figure 9). The Portage Escarpment, the eastern extension of the Allegheny Escarpment along Lake Erie, also displays high relative relief. The Bellefontaine Outlier provides the greatest relief found on the western Ohio plain. (Adapted from Smith, 1935)

The Flushing Escarpment in the far southeastern corner of Ohio, on the other hand, represents an old divide between major preglacial streams.

More localized areas of high relief have resulted from fortuitous variations in the rate of erosion. The Bellefontaine Outlier is our most notable example. The survival of this island of Devonian limestone and shale sitting on the Silurian dolomite plain of western Ohio has been attributed to its location as part of a preglacial watershed which escaped the vigorous erosion that removed Devonian strata from the rest of the plain. Its relative relief may have been increased by glacial scouring of the surrounding plain where the ice would have been thicker.

Other sources of regional relief are outcrops of resistant rock which maintain their elevations while erosion lowers softer strata surrounding them. The Black Hand Sandstone and the Sharon Formation are two important examples of resistant rocks that often occur as outcrops. Except where covered by thick glacial debris, surface exposures of these rocks form steep-sided uplands which give rise to fast flowing, actively eroding streams.

In contrast to the vast expanse of regional relief, local relief is associated with a single geologic feature. An example would be the cliff in which an arch is found. It is impossible to discuss the formation of a specific arch or pillar without also considering the local topography within which the feature is located. While regional topography may influence erosive power and put general size limitations on any arches and pillars that might form, local topography controls the amount of relief available at the site and thereby limits what finally appears there.

Both regional and local topographic relief is an expression of the erosional history of a site in which processes presently at work may have played only a minor role. Much of the existing relief of three-quarters of Ohio was greatly affected by continental glaciation, a force which is no longer active in the area. The erosional history of the unglaciated portion of the state is more complex, involving successive down-cuttings, rejuvenations and glacial diversion of drainage systems. Any consideration of the formation of geological features must take into account the possibility that conditions and forces quite unlike those presently found at the site may be responsible for much of what is now seen there.

Pathways for Water Movement

Although the presence of consolidated rock is a basic prerequisite for the formation of arches and pillars, that rock must not be so solid as to prevent water from moving through it if arches and pillars are to form. Permeability, the rate at which fluids can flow through a rock, is determined by the rock's porosity and the presence of openings.

Porosity is the ratio of the volume of void space to the total volume of the rock and is a function of the presence of small, often microscopic, interconnected cavities. Porosity can result from differential compaction of the original sediments, spotty cementation of the grains making up the rock or differential solution which removes some but not all of the rock fabric. Carbonate rocks often contain voids of various sizes. Clastic rock subjected to weathering can lose the cement holding its grains together which increases the porosity of surface exposures.

Not all rocks are porous, but even non-porous rocks can be permeable if their fabric is broken. The most commonly found discontinuities in sedimentary rocks are bedding planes, the generally horizontal divisions that result from changes in depositional conditions and separate individual layers in the rock (Figure 11). Water is often found moving as a broad sheet upon the bedding planes between these layers of rock, the amount of flow being determined by how open the bedding plane is and how much water is available.

Faults, which are fractures in the rock along which movement has taken place parallel to the surface of the break, can be important avenues of water movement into bedrock. Even if the fault does not open, it creates a weakness in the rock through which water can pass. Although there are faults in Ohio's bedrock, they do not appear to have had a major impact on the formation of the state's arches and pillars.

Joints, on the other hand, have been very important. Joints are generally vertical fractures along which there has been little or no

Figure 11. Horizontal bedding planes in Greenfield Dolomite, Paint Creek Lake Wildlife Area.

movement, and such movement as has occurred has been perpendicular to the joint faces, resulting in a wider opening (Figure 12). Joints are a reaction of rock to stress. Tension joints form to relieve elastic strain accompanying uplift. Joints in the coal beds of eastern Ohio, for instance, appear to have been caused by the compressive forces accompanying the rise of the Appalachian Mountains. Joints in flat-lying rocks that have undergone little disturbance may be fatigue cracks resulting from small, alternating stresses. The removal of overlying or surrounding rock by erosion can relieve the pressure rock strata are under, allowing the rock to expand and form joints which vaguely follow the configuration of the landscape. A similar process has been observed in deeply incised valleys after retreat of a glacier which filled it. Such off-loading has been responsible for forming joints paralleling the trend of gorges in Ohio and so has been important in the formation of many of the natural arches and pillars of the state. Glacial unloading has probably had much less effect since the continental glaciers covering the state were comparatively thin. They may

Figure 12. Vertical joint in Peebles Dolomite, Paint Creek Lake Wild-life Area.

have contributed to the widening of previously existing joints, how-ever, by sending vast floods of meltwater across the landscape.

Porosity, bedding planes and joints all provide avenues for wa-ter movement into and through the bedrock. Once there, water moves slowly and penetrates a large volume of rock, and so is able to accom-plish a great deal of chemical weathering. The interaction of these three rock characteristics with water determines the form which eroded features will take. Joints and bedding planes will be widened. Porous rock will have its pores enlarged as water moves through it, making it weaker and more likely to crumble. Cementing materi-als will be dissolved and removed, allowing more-resistant particles to fall away.

Where porous rock overlies impervious strata, water descend-ing through the overlying layers pools on top of the impervious rock and moves outward in every direction. If part of this underground flow reaches a valley where erosion has cut down to or below its level, the water emerges as a spring. This flow meets less resistance than does the flow in other directions and so captures a larger share of the available water. This focuses the erosive strength of the water into a narrower channel, enlarging the horizontal openings through which the water is flowing.

The same effect occurs in vertical crevices because they also offer water an energy-efficient route. As a result, vertical fractures can be enlarged. Such enlargement is especially prominent in calcareous rocks which are very soluble, but sandstones and conglomerates are also vulnerable. Many of Ohio's largest arches, such as Ladd Natural Bridge (Figure 144), were formed by the enlargement of vertical crevices and horizontal bedding planes.

Time

The last requirement for the formation of arches and pillars is adequate time. Geological processes move forward at a pace which is, by human standards, exceedingly slow. Weathering takes place molecule-by-molecule and grain-by-grain. Catastrophic events such as rock-falls which make large, abrupt changes in the landscape are rare. An arch or pillar usually represents the result of several millennia of erosion. At the same time, these features are, by geological standards, the work of a moment. Compared to the consolidation of sediments into rocks, the uplifting of mountains and the wearing away of continents, the appearance and disappearance of a natural arch or pillar takes place in an instant.

In the case of the arches and pillars of Ohio, that instant may be shorter than we at first think. During the Pleistocene, Ohio was invaded at least three times by massive sheets of ice moving down from the north. These continental glaciers began forming around 2 million years ago in the Arctic regions, and advanced and retreated several times in response to changing climatic conditions. Their great weight and the exaggerated effects of freezing and thawing they produced affected the bedrock they moved over, planing it off, grinding it down, pulling large pieces of it away, polishing and grooving it. In the end, they also covered it up. As the ice melted back in retreat after each advance, it dropped the load of ground-up bedrock it carried in a jumbled mass which in some places reached a depth of hundreds of feet. Ohio geologists examining this glacial "till" assign the oldest such deposits, found around Cincinnati, to the Kansan glaciation which occurred more than 240,000 years ago. Although Ohio was almost certainly affected by earlier glaciations, no certain evidence for them has been found here. Around 230,000 years ago the ice advanced once again. These Illinoian glaciers, named for the state in which their deposits were first studied, covered more of Ohio than any other continental glacier. At the height of the Illinoian glaciation, only the hilly southeastern corner of the state remained ice-free. By 125,000 years ago the Illinoian ice had melted out of Ohio and a more temperate interglacial period began. Then, 117,000 years ago, the latest glacial advance called the Wisconsinan began. Although this ice sheet did not cover quite as much territory as did the Illinoian, its effects are more noticeable since its deposits have been exposed to the forces of weathering for a much shorter time than have those of earlier glaciations. By 10,000 years ago the Wisconsinan glacier had also retreated and Ohio entered the present milder Holocene Epoch of the Quaternary Period.

Such delicate features as natural arches and pillars could not survive the tremendous weight and force of a continental glacier. Therefore, those arches found within the glaciated portion of the state must have begun forming after the retreat of the ice and so cannot be older than mid-Illinoian times. Most of them are even younger. One of the largest concentrations of natural arches in Ohio is found in the gorge of Baker Fork in Highland County which was formed by glacial diversion of previously existing drainage during the Illinoian glaciation. They therefore must have started forming after the retreat of the Illinoian ice. Arches found in gorges cut into the Niagara Escarpment of western Ohio by Wisconsinan meltwater streams are younger yet, as are those associated with wave-cut cliffs marking the various levels of Lake Erie and its predecessors.

The question of how old a particular arch or pillar might be becomes more complex for those found in the unglaciated southeastern third of the state. The various cycles of erosion which have affected this region have been incompletely studied and so the picture is still confused and the timing of these events is uncertain. In addition, continental glaciation also affected this area even though it was never covered by ice. Glacial meltwater rivers cut through it and much of the pre-existing drainage was rearranged. Many of the arches and pillars in the southeastern part of Ohio have formed in gorges cut by glacial diversion or by rejuvenated streams feeding into them. The study of the geologic features of unglaciated Ohio in relation to Tertiary Period stream systems and Pleistocene Epoch diversions holds great, and as yet unfulfilled, promise.

Chapter 2

About Natural Arches

Miller Natural Bridge (Figure 47) is what everyone expects a natural bridge to be. It is flat on top, very solid, spans an opening of respectable size and crosses water, at least when the intermittent stream beneath it is flowing. Having found it just where I was told it would be, I decided to walk the rest of the trails in this state nature preserve which I had not visited before. Rounding a corner and starting down into the gorge of Rocky Fork, I came face-to-face with a surprise — Miller Arch (Figure 48). While not as big as the natural bridge, it was in some ways more impressive given its location beside the trail and the suddenness with which it appeared. Later, when I asked my informant why he had not told me of the arch as well as the natural bridge, he replied, "Oh, that. I thought it was just a hole in the cliff."

Which, of course, is exactly what natural arches are — holes surrounded by rock. Unfortunately, geologic features rarely form discrete populations with well-defined boundaries. Natural arches are a good example inasmuch as they are a cluster within the much larger population of a common geologic feature most simply defined as "a hole in a rock." Such holes can take many forms. Which of them are counted as natural arches will depend on how "natural arch" is defined.

Definitions: A Sticky Wicket

Creating a definition which is clear, concise and comprehensive is no easy task. There is a surprising amount of variation in those features which most observers would intuitively call natural arches. In Ohio, Rockbridge (Figure 86) represents the classic version. It is flat-topped, beam-like in shape and crosses a valley. Camp

Christopher Natural Bridge (Figure 126) resembles Rockbridge only in having a flat top and crossing a valley, albeit a small one. In overall shape and manner of formation it is quite different. Miller Arch resembles neither and Rock House (Frontispiece) is like none of them. A proper definition must include all of these variations.

F. A. Barnes (1987) gives a concise overview of the quest for an all-inclusive definition of "natural arch" as it pertains to one of the most arch-riddled landscapes on Earth, the Colorado Plateau. Even here where the study of these features has been intensive and long-continued, no single definition has won complete acceptance. As Barnes points out, the terms "arch" and "natural bridge" were utilized by early observers because the few that were known at the time resembled the man-made structures after which they were named. Given the great variety in form found among the thousands of natural arches since discovered, many of which resemble neither arches nor bridges, a rigorous definition would begin by discarding the old terms in favor of words more specifically suited to the subject.

As it stands, however, the terms "arch" and "natural bridge" have become so ingrained in the literature and popular usage that any attempt to replace them would result in more confusion than the problem is worth. Furthermore, no adequate substitutes have been suggested. The wisest course appears to be to utilize these time-honored terms by giving them definitions specific to the topic at hand. We can start with the terms "arch" and "natural bridge" themselves. Although often used as synonyms, these two terms do not conjure up identical images. As early as 1910, Herdman F. Cleland attempted to discriminate between the two. In his report on the natural bridges of North America he stated that natural bridges cross valleys of erosion and natural arches do not. Most students of the topic since then have followed his lead, as will this report with one minor clarification. Here, "arch" will be considered the generic term for the features under discussion of which natural bridges comprise a specific subset. In other words, all natural bridges are arches, but not all arches are natural bridges.

Having settled on what to call these particular holes-in-the-rocks, we must now define them. Some authors have attempted to do so by referring to the human artifacts which a few of them resemble. Corgan and Parks (1979), for example, define "natural bridge" as "any natural feature, formed of rock, that resembles a manmade bridge." Thornbury (1969) defines a natural bridge as "a bridge or arch of natural rock." These and similar attempts say in effect that a natural bridge is a bridge that is natural. Hoffman (1985) gives the shortest definition. According to him, natural arches are "natural rock openings with unbroken spans." This, however, is too general, for it could be interpreted to include features such as alcoves which are not usually considered to be arches.

Resolving this issue of definitions is one of the major goals of NABS. To this end it has created a standards and definitions group which has proposed two definitions of a natural arch: a technical definition combining mathematics and geology suitable for the rigors of scientific research, and a "lay definition" for more popular usage.

NABS's proposed technical definition is as follows: "A natural arch or related type of natural rock opening is a rock exposure whose surface is described topologically as a 2-manifold in three space with one or more holes formed by the natural, selective removal of rock from a mass now represented by a relatively intact rock frame" (Horowitz, 1993). This definition is accompanied by eleven paragraphs precisely explaining the words used. It is based on topology, a branch of mathematics dealing with surfaces which considers doughnuts, coffee cups and natural arches to be one and the same.

The proposed lay definition is designed to be more easily understood by those of us without a Ph.D. in math: "A natural arch or related type of natural opening is a rock exposure with a hole completely through it, created by the natural removal of some of the rock to leave an intact rock frame around the hole" (Horowitz, 1993). This definition appears to include all the various geologic features considered to be natural arches. It also includes every solution cave with two or more entrances no matter how great the distance or twisting the course between them, and so may be too inclusive. It does, however, represent the best effort to date and has the added benefit of being sponsored by an organization attempting to coordinate the study of natural arches and bridges on a world-wide scale. It has therefore been adopted for use in this report.

Just as the technical definition requires an auxiliary explanation, so this lay definition needs clarification to ensure agreement on what is being said. "Rock exposure" refers to the parent bedrock of an area. "Natural removal" means that rock has been taken away by erosive agencies to form the hole. This and the requirement that arches have an intact rock frame eliminates gravity arches where the hole is formed by rock movement such as those found beneath isolated slabs of bedrock leaning against a cliff or between two slump blocks which happen to touch. On the other hand, slight deformation of an opening through movement along bedding planes or other lines of weakness caused by settling or similar natural causes does not disqualify a feature which otherwise meets the criteria established in the definition.

In some cases natural arches are found in rock that appears to have split from the parent bedrock and perhaps even moved slightly. In the Black Hand Sandstone region, slabs of rock of great size are often found separated from the cliff behind them by narrow passages which may have formed through erosive widening of a vertical crevice or through downhill creep of what amounts to a huge slump block. An arch formed in such a rock still qualifies for listing. The important consideration is whether or not the opening itself was initially created by rock removal through intact bedrock regardless of whether or not the bedrock containing it has moved. The hole must, however, go completely through the bedrock. Such a hole will have two entrances into the rock. In other words, it must be possible for something to enter the rock at one end of the hole and leave it at the other. The path it takes does not necessarily have to be straight or level, but it must be continuous.

The "intact rock frame" refers to rock with no open breaks. The frame may contain bedding planes or joints and still qualify as a natural arch so long as they have not been opened wide enough to allow something to pass completely through the frame and so out of the hole. If only part of a bedding plane or joint within the frame has been opened, but the remainder remains closed, then the rock is still intact and the arch qualifies under the definition, even if the joint or bedding plane can be traced through the frame.

Other terms used for natural arches are not as applicable to Ohio. Lohman (1975) regards arches as openings at or near the base of a rock wall. Openings found well above ground level are considered windows. No indication is given as to how far above ground level an arch must be to be called a window. McFarlan (1954) in writing of the natural arches of Kentucky refers to the small arches formed by the initial breakthrough of a ridge as "lighthouses." Both names seem to be a result of arches forming high above the normal viewing plane, allowing sky to show through their openings. There appears to be little benefit in this multiplication of terms.

On the other hand, the term "natural tunnel" is more useful. This is defined as a natural arch whose width is equal to or greater than three times its span (adapted from Corgan and Parks, 1979). Examples in Ohio include Rock House (Figure 104) and The Keyhole (Figure 39).

Terms Used in Describing Natural Arches

In developing the definitions of natural arch, natural bridge and natural tunnel, we have used terms which themselves require defining (Table 2). As a start, the words "natural arch" and "natural bridge" as used in this report will refer to the rock surrounding the hole rather than to the hole itself. "Natural tunnel," on the other hand, refers to the hole. Most of the terms used for the various components of a natural arch have been adapted from those established by NABS (Wilbur, 2005) in order to encourage the use of a common vocabulary. A few additional terms have been added here to aid the discussion.

Figure 13 represents an arch in its simplest form. As seen here, the defining feature of an arch — that which makes an arch of an otherwise solid exposure of bedrock — is the hole or void. This volume,

Table 2. Definitions of Natural Arches

Natural Arch	A rock exposure with a hole completely through it, created by the natural removal of some of the rock to leave an intact rock frame around the hole.
Natural Bridge	A natural arch which crosses a valley of erosion.
Natural Tunnel	A natural arch whose width is equal to or greater than three times its span.

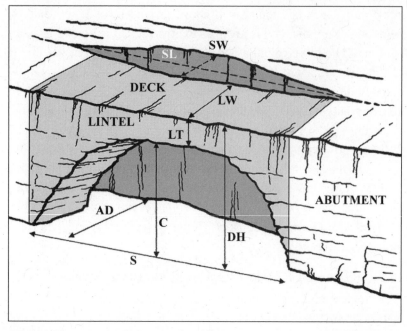

Figure 13. An arch in its simplest form showing the arch terminology used in this book: span – line S; clearance – line C; lintel width – line LW; lintel thickness – line LT; deck height – line DH; alcove depth – line AD; skylight length – line SL; skylight width – line SW. The lightly shaded area identifies the lintel, the extent of unsupported rock.

usually filled with air, is called the "opening." The two ends of the opening are its entrances. In a simple arch the opening is horizontal and both entrances occupy vertical planes. In more complex arches, the opening may have both a horizontal and a vertical component (Figure 14). In this report, this NABS definition of "entrance" is modified for the sake of clarity in discussing specific arches by limiting its use to those opening ends which are essentially vertical. Horizontal entrances will be referred to as "skylights," regardless of whether or not the sky is actually visible through them. In some arches the entrance and the skylight combine into a single, slanted passage, as, for example, in Boundary Arch (Figure 37); however, the entrance is still in the vertical plane and the skylight is still in the horizontal plane. In other cases, such as a small hole in the roof of a very wide alcove or

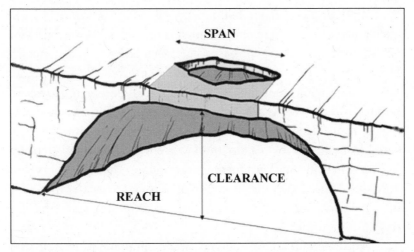

Figure 14. An arch formed by the creation of an opening in the roof of an alcove. The lighter shaded area represents the lintel and the darker shaded area represents the wall of the alcove. The opening of the arch has both a vertical entrance and a horizontal skylight. In this case, the span is determined by the length of the skylight and is shorter than the reach.

through a protruding horizontal rock layer, the opening may appear to consist of only the horizontal component or skylight. The vertical entrance is still present, but is so large in relation to the opening of the arch as to make its measurement of little use.

The characteristic of an arch which initially attracts our interest is its defiance of gravity; that is, the fact that the part over the opening has no support directly beneath it. Instead, its support comes from the points of its attachment to the parent bedrock and from the cohesive strength of the rock itself. The term used for this indirectly supported rock, indicated by the lightly shaded portion of Figure 13, is "lintel" which is defined as "the volume of rock in the rock frame that is unsupported except at two ends by the remainder of the rock frame." That part of the rock frame which supports the lintel is called the "abutment." Since by definition a lintel is supported at two opposing points, it naturally follows that it will have two abutments.

The base of the arch is the surface that is beneath the opening or lintel. In the strictest sense, this would consist of the bedrock forming the bottom of the opening, but in many cases this is covered by soil which may be difficult to remove or undesirable to disturb and so the NABS definition allows the base to be either bedrock or soil, but not loose boulders. It also allows the base to be water in those cases where the arch crosses part of a stream, lake or ocean. However, bodies of water have a habit of rising and falling which would result in differing figures for the height of the opening as measured from the bottom of the lintel to the water's surface, depending on when the measurements were made. There are very few arches in Ohio that could be considered to have a water base, and in every such instance it was possible to drop the measuring line to a solid bottom.

An important factor in comparing natural arches is their size. Since the opening is the primary characteristic of an arch, its dimensions are the main factor used in making such comparisons. The size of an opening is determined by its span which is the horizontal measure taken between the abutments, and its clearance which is the vertical measure taken between the bottom of the lintel and the base of the arch (Figure 13). While this might seem simple enough, it is complicated by the fact that natural arches are three-dimensional affairs. Span and clearance could be measured at either entrance or anywhere between them and the resulting figures could differ dramatically since openings rarely have walls that are perfectly straight and perpendicular. NABS guidelines suggest using the longest horizontal and vertical measurements in that part of the opening that is the most constricted, thus giving a true picture of the arch's defiance of gravity (Figure 15). A much more detailed explanation of how and why this is done can be found on the NABS web site, *www.naturalarches.org*.

One form of natural arch commonly found in Ohio is shown in Figure 14. Here a lintel has resulted from the removal of a small part of the roof of an alcove. In this case, the arch is part of a larger feature. The lintel itself, indicated by the shading, consists only of that segment of rock supported at two opposing points and includes only a portion of the outside edge of the alcove roof. While the width of the alcove entrance, here called the "reach," might appear to meet the definition of span and therefore be the rightful indicator of the relative

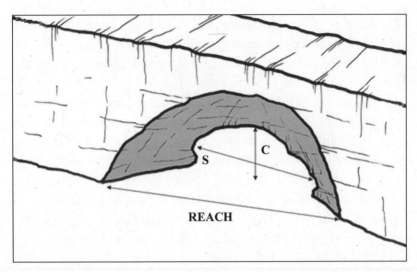

Figure 15. An arch in which the opening becomes smaller between the two entrances. The span (S) and clearance (C) are measured at the largest horizontal and vertical expanses of the most constricted part of the opening.

size of the arch, the rock surrounding the alcove is in fact supported along the entire base of the alcove wall rather than at two opposing points. The dimension designated "reach," the alcove width measured at floor level, may not be important in describing the arch itself, but may be useful in discussing features associated with it. The measure of reach varies across the depth of the alcove since alcove walls usually converge toward the rear. If not otherwise stated, reach is assumed to have been measured at the alcove entrance where it is usually the greatest.

In Figure 13, the opening through the roof of the alcove is wider than the opening of the arch. In this case, the back wall of the alcove, and therefore its supporting capacity, has been removed. The rock surrounding the remnant of the alcove is now supported at two opposing points and so meets the definition of natural arch. Its span is determined by the longest horizontal dimension of the most constricted part of the opening. In arches where the side walls of the opening parallel each other for their entire length, the span equals the reach. In Figure 13, therefore, the reach equals the span.

Several other terms will be useful in discussing natural arches (Figure 13). The top surface of the lintel which may or may not be flat will be called the "deck." The longest horizontal dimension at the narrowest part of the lintel measured perpendicular to the arch entrance is the "lintel width" which can usually be considered to equal the arch width. The plane of the arch facing down slope is considered to be its "front." In a natural bridge, this plane will be the one facing down the valley. In arches associated with cliffs and parallel to them, the front will be the plane facing away from the cliff. If the arch is perpendicular to the cliff, as in those formed in a fin of rock projecting from a cliff face, then the front will be the plane facing down the valley of which the cliff is a part. The plane of the arch facing upslope is its "back." In arches formed at the crest of divides, both planes face down slope. A similar situation might be found in arches resulting from the close proximity of two solution sinkholes with no water flowing through the underlying cave and therefore no indication of which direction is down slope. In such instances, both faces are by definition fronts, and so a different designation needs to be utilized. In these cases, compass terms such as "north face" and "south face" are used to differentiate between them.

The Question of Significance

There are literally millions of rock-bound holes found in Ohio that meet the definition of natural arch as given above. If a compilation of them is to have any meaning, some method of determining the significance of a given example must be devised. Significance can be measured in many ways. Vreeland (1994), one of the earliest and most far-ranging students of North American natural arches, used five criteria of significance: size, name, history, density (number of features in a given area) and appearance. Most other studies, including this one, consider size alone to be the only criterion of significance.

Upper size limits are not a problem. Landscape Arch in Utah, the longest natural arch so far recorded in North America, has a reported span of 290 feet (Wilbur, 2004). This is far larger than anything found in Ohio. Lower limits, on the other hand, are very much a problem. The number of features qualifying as arches rises exponentially toward the smaller end of the scale. Arches National Park in Utah, North America's premier collection of natural arches, contains only thirty significant arches if significance is defined as a minimum opening dimension of 50 feet. When the minimum is lowered to 3 feet, the number of significant arches jumps to 1374 (Stevens and McCarrick, 1988, 1991). This situation is not limited to Arches National Park. Rockbound holes are common geologic phenomena. If the minimum size requirement for listing those that are to be considered significant is set too low, the list can rapidly become unmanageable.

Setting a lower limit is strictly an arbitrary choice and varies among researchers depending on the number and relative sizes of the features under discussion in a given area. There appears to be no one point on the size scale at which the suite of arches naturally divides into significant and insignificant groups. The one ever-present, although usually overlooked, principle is that the chosen lower boundary must relate to the human scale since the recording is being performed by people. The list of significant natural arches would look quite different if it were being compiled by cockroaches.

Stevens and McCarrick's study of Arches National Park, one of the most intensive studies made of the arches of a particular region, considers an arch with an opening 3 feet or larger to be significant. None of the reasons given for choosing this lower limit concerned the arches themselves, but rather had to do with former studies and administrative conveniences. NABS considers any arch having one or more opening dimensions equal to 1 meter (3.28 feet) to be a "minor" arch. A "significant" arch contains two opening dimensions the product of which is at least 10 square meters (107.6 square feet).

For reasons given in the section on Methodology, adopting the 3-foot minimum for listing significant arches in Ohio seems justified. It should be noted that this size limitation applies to the opening through the rock. Stevens and McCarrick (1988) use a "rod test" in determining whether an opening qualifies. If a 3-foot-long rod held perpendicular to the arm can pass through the opening, then it is listed as a natural arch, provided it meets the other criteria as given in their definition. For the Ohio survey, no measured dimension of the opening can be less than 1 foot. Therefore, the smallest opening that would allow a natural arch to qualify for listing in Ohio is 3 feet by 1 foot. This means that a number of arches considered significant for the Ohio survey do not qualify as such under NABS standards.

Measuring Ohio's Natural Arches

When comparing geologic features, physical dimensions are one of their more useful characteristics. Physical dimensions enable us to describe the features more completely for purposes of analysis, to rank them according to various traits (longest, highest, etc.) and to assign them a place in various systems of significance based on size. In order for such comparisons to have value, the same dimensions must be measured for each feature and the measurements must be repeatable within reasonable limits.

In the case of natural arches, the most common dimension measured is length. What most tabulators of arch statistics appear to mean by this term is the length of the lintel — that is, the distance between the two opposing points of support. The term "length" is a bit ambiguous, however, since it implies that it is the greatest of the measured dimensions, and in many arches this is not true. For that reason, the term preferred in this report is "span."

The measurement of an arch's span is often complicated by the shape of the opening forming it. The actual characteristic which needs to be indicated by this measurement is the arch's "defiance of gravity" — that is, the length of unsupported bedrock forming the arch. Since the length of unsupported bedrock is directly related to the vertical pull of gravity, only the longest horizontal expanse of the opening can be used to determine the span of an arch. In cases where the base of one arch abutment is higher than the other, the length of the line between them will not give a true picture of the arch's defiance of gravity. This problem can be overcome by defining the measure of span as the length of the horizontal component of the opening regardless of its linear orientation. In such cases, the measured length of the span may be shorter than the opening length which is defined as the longest dimension of the opening. In Figure 16, the longest horizontal expanse (span) of the middle opening equals the longest expanse

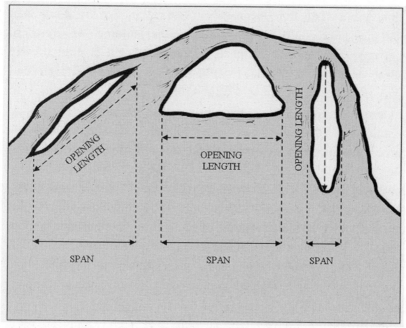

Figure 16. Span and opening length. Although all three openings in this diagram have approximately the same length, their spans, which are limited to the maximum horizontal distance between the abutments, vary greatly.

of the opening (opening length). In the opening on the right, the opening length is perpendicular to the span which equals the opening width (opening dimension perpendicular to the opening length). The opening on the left side illustrates an intermediate state of affairs. A very irregular opening may require that the span be measured in sections.

Almost as important as the length of the span in defining an arch is its clearance — the vertical distance from the bottom of the lintel to the base of the arch (Figure 13, line C). When the arch opening has no apparent base, as in Raven Rock Arch (Figure 117), clearance is measured from the bottom of the lintel to the nearest approach of bedrock below.

Unfortunately, measuring span and clearance is more complicated than this simplified discussion may indicate. Whereas it treats openings as two-dimensional shapes, arch openings are, in fact, three dimensional. It is not unusual to find large variations in the dimensions of an arch opening from one entrance to the other. Span and clearance could be measured at either entrance or anywhere between them and the resulting figures might differ dramatically since openings rarely have walls that are perfectly straight and perpendicular. Hole-in-the-Wall Arch (Figure 130) provides an extreme example of such a situation. Here, two alcoves intersect, resulting in a small opening between them. In passing through this arch, one finds a wide entrance narrowing to the small hole broken through the common back wall of the alcoves, and then widening again toward the entrance of the second alcove. In such a case, NABS guidelines suggest using the longest horizontal and vertical measurements of that part of the opening that is the most constricted, thus giving a true picture of the arch's defiance of gravity (Wilbur, 2005). The longest horizontal dimension of this part of the opening is the span of the arch, and the longest vertical dimension is its clearance (Figure 15).

In pillared-alcove arches such as 1811 Arch (Figure 107) this tidy solution is deranged. In such an arch the two entrances to the opening both occupy vertical planes, but are located side by side, separated by a pillar which splits what would normally be a single alcove entrance into two. The opening of the resulting arch is shaped like a U lying on its side and curving around the pillar which in effect becomes a vertical lintel. In this case, clearance, which is defined as a vertical measure, is the height of the pillar taken on its interior side. Span is the measure of the horizontal distance between the interior face of the pillar and the farthest extent of the alcove, equal to alcove depth minus the width of the pillar. By imagining a pillared-alcove arch rotated ninety degrees so that its vertical pillar becomes a horizontal lintel, one can see that the two dimensions measured are comparable to the span and clearance of a normal arch, only reversed. While this method of measuring pillared-alcove arches may appear to give an exaggerated picture of its span, it fits well with the system used for the other arch varieties and eliminates any confusion over what actually was measured.

Another complication occurs with arches having more than two entrances. Rock House, for instance, has an entrance at each end of its opening, as would be expected; it also has five entrances through the side wall. Is it one arch or four, or maybe even seven? Rock House is not an isolated example. In some cases, such as Jawbone Arch where a vertical crevice has widened in two places to form skylights, both of which open into the same long alcove, it seems reasonable to consider the skylights as two separate arches. Rock House, on the other hand, has always been thought of as a single feature, having one main opening with an entrance at each end. The other entries are considered to be breaks in the side wall. Following this example, similar multi-entrance arches are also considered to be single features. In the end, it must be admitted that the great variety of these features makes it difficult to establish a hard-and-fast rule for determining whether multi-entrance arches should be counted as one arch or as multiple arches. Each situation must be considered on its own merits.

Dimensions of the lintel are also important. Lintel width is the horizontal dimension perpendicular to the span; lintel thickness is the vertical dimension perpendicular to the span. Both measurements are made at the most constricted part of the lintel. Where the lintel is not horizontal, as in pillared alcoves, the terms "lintel breadth" and "lintel depth" replace "lintel width" and "lintel thickness." A lintel can vary dramatically in width and thickness from one end to the other. These variations are noted when they are important factors in describing an arch.

Measurement of the skylight, when present, can also be useful. Skylights may be narrow crevices or wide, bowl-shaped openings. Skylight length is the longest horizontal measure of the opening parallel to the span. Skylight width is the longest horizontal measure perpendicular to the span. Here we run into the problem of the terms themselves implying a certain result in relation to each other which may not always exist. It is not unusual for skylight lengths to be shorter than their widths.

Although not an essential measurement, "deck height" (Figure 13, Line DH) defined as the longest vertical distance between the top of the lintel and the bedrock beneath it, is often of interest for descriptive purposes. Deck height measured at the front of an arch often differs from that measured at its back. Unless stated otherwise, deck

53

height refers to that found at the front of the arch. This is usually the most impressive face of the arch and is often the one recorded in photographs and drawings.

The amount of penetration into bedrock of an alcove from which an arch is formed may be useful in giving a more complete picture of the feature under discussion, especially in those cases where the alcove or alcove remnant extends into the rock beyond the arch and the skylight. The term "alcove depth" designates this measurement and is defined as the longest horizontal line between the entrance of the original alcove and its farthest point of penetration.

These definitions provide a foundation for the measurement of natural arches. Nature, however, is not so predictable. Each arch is unique and requires individualized treatment. Additional measurements, important variations and further explanations are utilized as needed in the arch descriptions in the catalog to provide as complete a picture of each feature as possible.

How Natural Arches Form

There are many places where all five basic conditions required for the formation of arches are present, but no arch forms. Not only must the conditions be present, but they must also interact in specific ways. It is these interactions which explain how a given arch forms. In some cases the manner of formation appears obvious; in others, it is more difficult to determine. Since it is rarely possible for a human observer to watch an arch form from beginning to end, any explanation of the process is of necessity an interpretation of the evidence at hand, a best guess limited by the abilities of the observer and the state of knowledge at the time. By taking careful note of the surroundings of an arch and ferreting out such clues as are available, an investigator can make a reasonable assumption as to the specific processes involved in its formation. Even so, a good bit of art is necessarily mixed with the science. This means that interpretations of the same feature can vary between observers and even with the same observer over time as more details are discovered.

A good example of this is found in reports on the famous Natural Bridge of Virginia. Known and studied since the days of colonial

America, this large arch which includes George Washington and Thomas Jefferson as former owners has been described by different authors at different times as the remnant of a collapsed cave, a result of waterfall capture, an undercut stream-meander spur and evidence of stream piracy through underground channels. That such a well-known and long-studied feature should have so many plausible explanations emphasizes the challenge inherent in interpreting them.

Some of the difficulty of interpretation can be attributed to past events. Ice sheets wax and wane, sea levels rise and fall, precipitation patterns shift. The processes which are now having the most effect upon the land may not have been important in the past. Other processes not apparent now may have been of far greater importance in creating the landforms we are trying to explain. To further complicate matters, the present form of a natural arch may have resulted from the interaction of several different processes. For example, the skylight of a natural bridge formed by crevice enlargement at the lip of a waterfall might be enlarged by the collapse of roof blocks into an underlying alcove. Every geologic feature has its own particular history.

In spite of this great variety, however, natural arches do exhibit enough similarities in both appearance and apparent manner of formation to allow the creation of an orderly system of study. Several such systems have been devised, the most elaborate being that proposed by NABS (Wilbur, 2005). The system used here is based on how the openings were formed. It was adopted because it gives a clear, comprehensive framework into which the arches found in Ohio readily fit.

All natural arches listed in this report are a result of the erosion of zones of weakness in bedrock. The manner in which this is accomplished varies from site to site, but enough similarities exist to allow us to recognize six categories of arch formation in Ohio: breached alcove, enlarged vertical crevice, enlarged joint and bedding planes, bedrock texture remnant, opened bedding planes and collapsed cave. These categories reflect the geological conditions responsible for arch openings; in other words, they state what the arch actually is. Secondary characteristics such as location in the landscape and shape of the lintel which have been used in other systems to define arch types are here considered to be descriptors which may segregate a collection of arches that cuts across category boundaries. For example, in this system an arch which has formed through the enlargement of a vertical crevice is an enlarged-vertical-crevice arch regardless of whether it was widened on dry land by atmospheric erosion or on a coast by wave action. The latter case in most systems would be considered a coastal (or sea) arch. In the Ohio system, it is an enlarged-vertical-crevice coastal arch. There can also be breeched-alcove coastal arches, enlarged-joint-and-bedding-plane coastal arches, bedrock-texture-remnant coastal arches, opened-bedding-plane coastal arches and perhaps even collapsed-cave coastal arches.

While broad generalizations can be helpful in organizing natural arches into categories, the same variations which make defining them so difficult ensure that they will rarely fit cleanly into any categorical listing. Nor must we ignore the very real possibility that the store of knowledge from which this system, or any system, is built may be, and probably is, woefully incomplete. A sense of humility in the presence of nature's wondrous creativity is a valuable asset. It is with a full awareness of these limitations that the following categories of arch formation are offered. At this point, they appear to fully account for the natural arches presently reported in Ohio. No claim is made for their usefulness in other regions, and the right to add to, subtract from or change them completely as more information appears is fully reserved.

Categories of Natural Arch Formation in Ohio

Breached Alcove

Breached-alcove arches result when the ceiling or wall of an alcove is broken through in such a manner that part of the alcove remains intact, forming a lintel (Figure 17). The breaching can be a result of collapse, joint widening or atmospheric erosion.

Some of the largest natural arches in Ohio have resulted from the partial collapse of alcove roofs where the outermost rim remains to form the arch. In these cases, intersecting vertical joints in the ceiling have been widened enough to isolate blocks of rock which then fall. Skylights caused by roof collapse tend to be angular in outline with straight, joint-determined sides.

Figure 17. Lucas Run Natural Bridge, a breached-alcove arch.

Another common means of alcove breaching is by the widening of a joint parallel to the face of the cliff in which the alcove has formed. When such a joint is opened by weathering, the band of rock remaining between it and the front of the alcove becomes a natural arch. The alcove provides the entrance and the opening of the arch, and the widened joint becomes the skylight. The lintel parallels the cliff face, often forming a continuation of it, and appears blocky and angular with a relatively flat top. Skylights in such arches are usually elongated parallel to the span and often narrow into the original crevice at each end. Miller Natural Bridge, shown in Figure 47, is an example of this.

Alcoves can also be breached by growing too large for their surroundings, breaking into a pre-existing space. Alcoves enlarging upwards can break through their ceilings. This is often the result of an enlarging tension dome (Corgan and Parks, 1979). These domes form in layered rocks over an opening, be it an arch or a cavern (Figure 18). Each layer of rock acts like a beam supported only at the ends. Under the effect of gravity, the rock beams sag and eventually collapse into the opening. The entire rock beam does not fall, however. Only the center of the beam where the sag is greatest fails. The ends of the beam anchored into the bedrock remain and give added support to the rock beam above it. When this higher beam finally collapses, a shorter length fails. As higher beams or layers collapse into the opening below, ever smaller segments of rock are involved until the gap is so small that the strength of the rock is great enough to span it without collapsing. The result is an opening with a sloping top, similar to that found in certain Mayan buildings of Central America where openings were capped with ever longer blocks of rock. In a cave chamber or alcove, this results in a dome-shaped ceiling (Figure 19). In an arch, the result is an opening with a semi-circular top (Figure 20). The man-made "natural arch" in Castalia Quarry

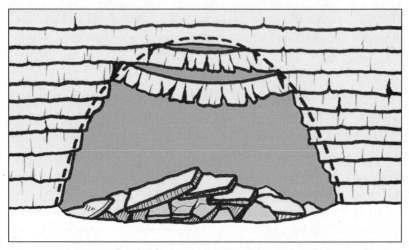

Figure 18. Steps in the formation of a tension dome. Rock layers sag and collapse into a void. Higher layers of bedrock still in place are partially supported by lower layers so shorter lengths of bedrock collapse, causing the void to enlarge into a dome-shaped cavity. The extent of the tension dome is marked by the broken line. (Adapted from Corgan and Parks, 1979)

Figure 19. A tension dome formed in the roof of the alcove at Early Arch.

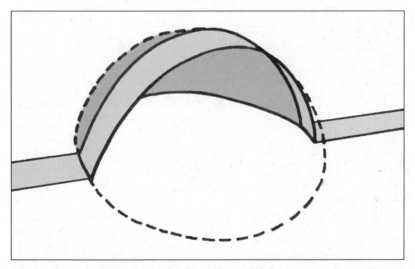

Figure 20. A tension dome forming in a cliff face results in an alcove, shown by the darker shading. If the dome forms in a narrow wall of rock, or the back part of the alcove is removed by erosion, an arch can result, as shown here by the lighter shading.

Reserve, shown in Figure 21, clearly illustrates this process. Although rock within a tension dome may be in danger of falling, the dome itself is one of the most stable methods of covering a void. That is why the dome and its related forms, the alcove and the arch, appear so often in both nature and architecture.

Alcoves can also enlarge laterally. An alcove located in a narrow fin of rock might eventually break through it completely, leaving the remnant of the fin as an arch. Headward-erosion arches can form where the upper reaches of two streams eroding into a steep-walled ridge from opposite sides create opposing alcoves which enlarge until they connect. It is not unusual for the alcoves to be quite large in comparison to the small streams leading away from them. Headward-erosion arches most commonly form in high, narrow walls composed of strongly jointed rocks and are common in Kentucky's Red River Gorge region. Ohio's largest example of a headward-erosion arch is Pike Arch (Figure 128) where connecting alcoves have been eroded

Figure 21. The upper portion of the man-made "natural" arch at Castalia Quarry Reserve illustrates the formation of a tension dome through the failure of successively shorter lengths of rock layers.

by two intermittent streams into opposite sides of a projection of sandstone. More common here are arches formed by the uniting of adjacent alcoves which join through their side walls, usually resulting in pillared-alcove arches such as Early Arch (Figure 108).

Alcove enlargement arches can also form in high-energy shoreline regions of oceans and large lakes where waves cut alcoves, often called "sea caves," into narrow bedrock fins. If opposing alcoves are cut on either side of the fin, they may enlarge enough to connect through their common back wall. If the rock projection is thin enough, a single sea cave might be deepened enough to penetrate it. Amherst Arch North is an Ohio example, although the coastal environment which created it has long since vanished (Figure 113).

A special type of alcove enlargement arch important in some areas is the meander arch. Rivers in the mature stage of development tend to wander across the landscape, forming meandering loops. Should the region be uplifted, the river may be rejuvenated, gaining increased cutting power and entrenching itself into the bedrock. If the meandering course of the river is preserved during the entrenching process, the rock surrounded by a loop of the stream may take the form of a knob connected to the main mass of bedrock by a tall, narrow fin of rock. Since the erosive force of a river is strongest on the outside edge of a curve, a river entering an entrenched meander will throw most of its strength against the base of the fin. As the river flows around the knob at the end of the meander, its force shifts to the outside wall. Then, as the river moves into the downstream portion of the meander, its force shifts once more, again striking the base of the fin. Under these conditions, alcoves often form on opposite sides of the narrow neck of the meander. If the bedrock at river level is relatively weak and the overlying bedrock is strong, the alcoves may enlarge enough to join beneath the cap rock, creating an arch. Rainbow Bridge in Arizona, one of the largest natural arches in the world, formed in this way. The only possible example of a meander arch found in Ohio is Trimmer Arch which formed in a fin of dolomite left at the junction of two small streams (Figure 66). Any evidence of meanders or alcoves that might have existed has disappeared, however, and in its present form it appears to be a result of opened bedding planes.

A rare means of alcove breaching is through the enlargement of a stream-cut pothole. Chapel Ridge Arch is the only example of this process so far found in Ohio.

In breached-alcove arches the original form of the alcove is usually obvious. The manner in which the breeching occurred may be less so, but is usually apparent with a little study. The recognition of breached-alcove arches can be complicated by excessive enlargement of the skylight or retreat of the cliff face in which the arch is found. While it may seem unlikely that the solid bedrock of a cliff would weather away faster than the apparently fragile rock forming an arch, such is often the case. The resolution of this apparent contradiction may lie in the realm of groundwater circulation. Groundwater passing through the openings of bedrock serves as a major agent of chemical weathering. Rock penetrated by groundwater is gradually weakened and so made more vulnerable to erosion. When the rock frame of an arch is isolated, groundwater circulation through it is cut off. It is no longer subjected to the same rates of chemical weathering as the intact bedrock behind or beside it. Also, since the arch is exposed to open air on several sides, sunlight and wind can rapidly dry it, unlike the intact cliff which presents only one face to the air. Since the effectiveness of weathering advances with increased contact time between rock and moisture, it follows that the intact bedrock will erode at a faster rate than will the isolated arch.

Cascade Park in Elyria protects an interesting example of a breeched-alcove arch in-the-making. Here a steep-walled rock barrier 25 feet high and barely 30 feet wide locally called the "Natural Bridge" juts out from the western wall of the Black River Gorge, ending at a small, shale-topped hill called the Camels Back. The north rim of this rock wall was at one time the lip of a large waterfall formed by the West Branch of Black River. The force of this fall carved out a circular plunge pool at its base which remains today as the Amphitheater. It also hollowed out a low but deep alcove in the rock behind it. At this time, the East Branch also formed a falls on the same line, the two waterfalls being separated by the Camels Back. The eastern falls must have carried more water than the western falls or conditions for erosion were more favorable there, for it cut its way upstream faster.

This resulted in the flow of the West Branch being diverted from the falls to meet the East Branch as it does now. In cutting down to keep pace with the East Branch, the West Branch isolated the lip of its former falls, creating the wall now called the Natural Bridge. In doing so, it almost cut into the rear of the alcove eroded at the base of the former waterfall. A small or very determined person crawling into this alcove realizes that only a few feet separate its back wall from the gorge of the West Branch. Had the breach been completed, this would have been one of Ohio's more impressive natural arches.

Enlarged Vertical Crevice

The enlarged-vertical-crevice category of arch is formed by the widening of a vertical crevice that lies perpendicular to the cliff face (Figure 22). It opens into a skylight which may be a vertical-sided pit as in Cats Den Natural Bridge (Figure 125), an interior valley as in The Keyhole or the top of the crevice itself if it breaks through to the surface behind the span as in Boundary Arch (Figure 37).

Enlarged-vertical-crevice arches have strongly vertical openings which may narrow to an impassable crack or pinch out altogether at the top. The widened part of the crevice often displays a complex silhouette produced by differential erosion of the beds making up its walls. The floor of the opening may be horizontal or it may slope steeply up from the entrance to meet the skylight.

Enlarged Joint and Bedding Plane

The combination of vertical joints paralleling a cliff and horizontal bedding planes or a less-resistant layer of rock intersecting both is a very common occurrence in some rock types. This combination of circumstances can be brought together in several ways. A stream carving a gorge may be guided by joints already cutting the bedrock. Since multiple joints are often found paralleling each other, those not used by the stream would naturally lie parallel to both its course and the walls of the gorge it cuts. Joints parallel to a cliff face can also result from off-loading after formation of the gorge. Bedrock is under stress, both horizontally and vertically, from the weight of rock above and on each side. When that weight is removed through erosion, the

Figure 22. The Keyhole, an enlarged-vertical-crevice arch.

bedrock often expands in the direction of removal to relieve the stress. Due to the relative brittleness of rock at the surface of Earth, this expansion can result in the formation of cracks or joints paralleling the exposed surface. In the case of a river-cut gorge, the stress will be relieved into the gorge and the resulting joints will parallel the cliffs forming its sides.

Bedding planes, the other half of this equation, are ubiquitous in the sedimentary rocks exposed at the surface of Ohio, and resistant layers of rock lying above less-resistant layers are quite common. There are a great many opportunities for these to intersect with vertical joints. Given the large number of these meetings which exist, it is not surprising that some of them would be so situated as to give rise to natural arches (Figure 23).

In most of the Ohio examples of arches formed through enlargement of a joint and bedding plane intersection, a small, usually intermittent stream drops into a skylight and flows out beneath the arch. This stream is often responsible for opening the arch. The original flow of such a stream would have been over the edge of the cliff, forming a waterfall. Some of the flow, however, seeps down into vertical joints crossing its bed. If this seepage reaches a point where continued downward movement is blocked by a resistant layer of rock, the water will pool, spreading out between bedding planes and filling

Figure 23. Ladd Natural Bridge, an enlarged-joint-and-bedding-plane arch.

pores within the bedrock. When this expanding pool reaches the cliff face, it finds an outlet. A flow is then set up which pirates water from the stream, channeling it down behind the original falls and out its face. Here it may initiate formation of an alcove or help to enlarge one created by the sapping action of the waterfall. This pirated flow widens the joint, allowing it to capture more of the stream which in turn accelerates its widening and further enlarges the alcove. This erosion may continue until all of the stream's flow is channeled through the enlarged openings, leaving the rock lip of the original falls as a natural bridge. Continued weathering and erosion by the pirated stream enlarges the skylight and arch opening. In many cases, the stream has eroded a notch in the back wall of the skylight. Some of Ohio's most spectacular arches have formed in this way. Several waterfalls exhibit the early stages of this method of arch formation. It is reported that during periods of low flow, the water of Hayden Falls on Hayden Run, a small tributary of the Scioto River in Franklin County, ceases to go over the lip of the falls, dropping instead into cracks some distance above it and coming out the face of the falls (Stauffer, Hubbard and Bownocker, 1911). Here we have a natural bridge under construction, requiring only continued enlargement of the crevice and bedding planes now being used by the water of the stream.

The alcoves of enlarged-joint-and-bedding-plane arches, especially the waterfall type, can become quite large and result in their being mistaken for breeched-alcove arches. The difference lies in the relationship of the alcove to the vertical crevice. In breeched-alcove arches, the alcove formed first and was broken into at a later time. The alcove usually extends into the bedrock beyond the breech, and by its size and shape, this extension proves itself to be an obvious continuation of the original alcove. In joint-and-bedding plane arches, on the other hand, the joint was enlarged at the same time that the alcove was being formed between it and the cliff face. The alcove usually ends at the straight back wall of the joint forming the skylight. In those cases where a secondary alcove has been eroded into this wall, it will be smaller than the main alcove or show in other ways that it has been formed more recently, either by sapping action of the waterfall plunging through the skylight or by groundwater seeping through bedding planes.

Bedrock Texture Remnant

Bedrock texture is the distinct fabric imparted to the landscape by characteristics of the bedrock which forms it, as for example, cross-bedding in sandstone, the orientation of fossils in rock strata and zones of easily eroded rock surrounded by harder rock (Corgan and Parks, 1979). Every layer of bedrock exhibits variations. Grain size and orientation, amount and type of cementing material, porosity and a number of other characteristics can change as a rock layer is traced vertically and horizontally. As the rock changes, its resistance to weathering also changes. This variation often results in masses of comparatively harder rock intermixed with softer rock. These masses of hard and soft rock are not necessarily monolithic in nature, but may be invaded and even penetrated by each other. There may be no visible sign of these variations in fresh exposures, but weathering makes them obvious by removing the weaker rock, leaving the harder rock behind. As the harder lumps of rock emerge, they may take the form of natural arches. Cedar Fork Arch (Figure 24) is an example.

Several of the dolomite arches found in Highland and Adams counties appear to have formed in this way. They have a ragged and irregular form with deeply pitted surfaces, and openings that are more sinuous than those found in other categories of arches, often appearing as simply larger examples of the smaller pits that honeycomb their surface. Removal of the less-resistant rock which once surrounded them may have been aided by circulating groundwater which carried away the soluble material, leaving only insoluble residues to be removed by surface erosion from around the arch.

Bedrock-texture arches formed in sandstone or conglomerate often show a distinct relationship to cross-bedding found in the parent formation. One or more surfaces of the arch may follow in a general way the contours defined by the sloping beds. Such arches may appear as hollowed-out places where softer rock has been removed. This process is most likely the cause of many pillared-alcove-type arches in which both entrances to the void occupy nearly the same

Figure 24. Cedar Fork Arch, a bedrock-texture-remnant arch.

vertical plane. Although alcove enlargement can also form pillared-alcove arches, the large size of the pillar in comparison to the alcoves and the still-apparent dual nature of the alcoves differentiates this type from those formed through bedrock texture in which the pillar is usually thin in relation to the obviously single alcove behind it.

Because the erosion responsible for such arches usually takes the form of atmospheric weathering, their formation can be expected to take longer than is the case where more rapid forms of erosion are involved. All of the known arches formed in this manner are found either in the unglaciated part of the state or in areas subjected only to Illinoian glaciation, regions of Ohio where erosion processes have had the most time to work without subsequent alteration by glacial activity.

The only exception to this is natural arches found underground (Wilbur, 2007). Cave walls often appear to have been carved into scallops and flutes. Fins and shelves of rock may be pierced by openings

Figure 25. Olentangy Caverns Arch, an underground bedrock-texture-remnant arch.

which can reach respectable size (Figure 25). Some of these openings may qualify for listing as natural arches. Most of them appear to have been formed through differential solution of the rock and so would be considered bedrock-texture-remnant arches. The only examples of this process noted so far in Ohio are found in Olentangy Indian Caverns which is located in the glaciated part of the state.

Opened Bedding Plane

Some arches are a result of the enlargement of openings along bedding planes, usually created through a combination of chemical solution and mechanical breaking. Ohio's examples of this type of arch occupy narrow projections of rock formed of a massive upper layer of bedrock and a weaker lower layer with many closely spaced bedding planes. The openings of these arches have been made through the lower layers.

Bedding planes may be horizontal or they may be angled. Tecumseh Arch (Figure 26) has formed in rock with horizontal bedding planes. Miller Arch (Figure 48), on the other hand, has formed in a localized section of Peebles Dolomite which exhibits arched bedding planes. The means by which the openings in such arches have formed is not always apparent. It may be a result of stream erosion during an earlier and higher phase of valley cutting. The usual absence of rock debris in the opening would appear to favor this explanation. Then again, the opening might be the result of atmospheric erosion processes. This would require some sort of localized weakness in the rock which could be exploited by weathering to form the opening; otherwise, the rock forming the abutments would have been

Figure 26. Tecumseh Arch, an opened-bedding-plane arch.

removed at a rate equal to that of the rock where the opening now is and no arch would have formed.

Collapsed Cave

Collapsed-cave arches could be considered a variation of breached-alcove arches in that both are formed by the failure of roof rock over an opening in bedrock (Figure 27). However, in cave-collapse arches the original opening is much deeper than an alcove. Caves are defined as natural cavities or a series of chambers excavated in bedrock beneath the surface of the earth (Howell, 1960). Most caves are formed by solution of carbonate rocks, usually limestone or dolomite, by the action of groundwater. Technically, such cavities are considered true caves only if they extend beyond the reach of external light coming through any opening to the surface. True caves can have lengths exceeding 100 miles; Mammoth Cave and Carlsbad Caverns are famous examples of such lengthy cave systems. This strict definition, however, eliminates many smaller cavities and cave remnants which obviously formed in the same manner — a consideration of some importance in Ohio where "true" caves are both rare and small.

Figure 27. Tiffin Arch, a collapsed-cave arch. Part of the collapsed roof is visible to the left of the arch opening.

If a cave enlarges to the point where its roof can no longer support its own weight, collapse occurs. On the surface, the failure may produce a pit called a sinkhole, often in the shape of a funnel which may or may not open into passages which can be entered by human explorers. If two sinkholes form close to each other and the section of cave roof remaining between them has an open passage beneath it, a natural arch results. Because caves in Ohio are small and few, cave collapse arches also will be small and rare.

Methods of Formation of Non-listed Arches in Ohio

Scattered across the landscape are thousands of arch-like forms which do not meet the definition of "natural arch" as used in this report. Many of them are interesting in their own right and some are quite picturesque, even spectacular. Although none of them are part of the official list, they deserve consideration as being an important part of the geological story. As with listed natural arches, arch-like forms can result from a variety of processes.

Gravity Arches

By far the greatest number of arch-like forms results from rock movement and can be grouped under the general heading of gravity arches. Unlike the openings of listed arches which form through the erosion of intact bedrock, the openings of gravity arches are a result of movement by the surrounding rock. This movement can take place in a number of ways. Slump blocks during their migration downslope can come to rest against each other or against a cliff in such a manner as to form an arch. Sulphur Creek Gravity Arch in Fort Hill State Memorial, Highland County (Figure 28), has formed from two slump blocks leaning against each other. A small stream flowing through the opening has widened it and given it a profile similar to that of a true natural arch. Rocky Fork Needles Eye in Rocky Fork Gorge, Highland County (Figure 29), is a more dramatic example. Gigantic slump blocks clog the narrowest section of the gorge. Two of them lean against each other to form Needles Eye. Most of the flow of Rocky Fork passes through this opening and has widened and smoothed it. Passing through this impressive arch is one of the highlights of a canoe trip on this rock-walled river. Slump block gravity arches can be

Figure 28. Sulphur Creek Gravity Arch, formed by the contact of two slump blocks, at Fort Hill State Memorial.

Figure 29. Rocky Fork Needles Eye, a large example of a gravity arch. Much of the flow of Rocky Fork passes through this opening.

quite large. Slab Arch (Figure 30) formed when a large tabular piece of dolomite was undercut and collapsed, rotating downward and coming to rest against the cliff.

Gravity arches can also form through creep. Gravity pulling fractured rocks apart can produce long, narrow openings which may become gravity arches if slippage occurs along bedding planes in such a way that a layer of rock remains in place over the opening. The example shown in Figure 31 resulted when a slump block fractured and the smaller piece tilted outward beneath the unbroken and more stable cap rock.

Another type of creep, found where massive permeable beds overlie less permeable layers, is responsible for some of our largest gravity arches. Seepage along the base of a cliff formed in such rocks can destabilize the bottom of an overlying block which has been separated from the cliff by weathering along vertical fractures. If the

Figure 30. Slab Arch, located in Fort Hill State Memorial, is a gravity arch formed of a slab of dolomite which was undercut by erosion and then collapsed.

Figure 31. This gravity arch in Clifton Gorge State Nature Preserve formed when slippage along bedding planes widened a vertical fracture beneath the caprock of a slump block.

bedding planes have a slight downward slope away from the cliff, the base of the block may slide outward while its top remains in place against the bedrock behind it. The resulting opening will be vaguely triangular in silhouette (Figure 32). These gravity arches often appear to meet the definition of natural arch required for listing, but can be identified by an often slight difference in bedding plane angle between the block and the cliff, and by an equally slight difference in elevation between the rim of the cliff and the top of the lowered block. Such arches are commonly found in the Hocking Hills and in the Sharon Formation "ledges" of northeastern Ohio.

Less common are gravity arches formed by the sagging and breaking of large, tabular slump blocks (Figure 33). These can be

Figure 32. A gravity arch in John Bryan State Park formed by the outward creep of the bottom of a large slump block while its top remained lodged against the cliff behind it.

recognized by the opposing tilt of the two pieces of the slump block and by the triangular profile of the opening, often with matching projections and recesses on each side showing where the different layers making up the block broke apart.

Gravity arches also form when boulders become wedged in the upper part of widened crevices, forming a short roof (Figure 34). These lodged-boulder arches, sometimes called chock-block arches, are usually small and their manner of formation very obvious.

Most gravity arches are readily recognizable as such. In some, however, atmospheric erosion has blurred the evidence of original rock movement. Determining whether or not a particular rock-bound opening is a gravity arch will depend on studying the overall setting

Figure 33. An unusual gravity arch in Clifton Gorge State Nature Preserve which formed when a long, tabular slump block sagged and broke.

Figure 34. A lodged-boulder gravity arch.

of the arch rather than concentrating only on its opening. Noting any differences in bedding plane angles in the various parts of the arch as well as breaks in the lintel will also be helpful in determining if an arch has been formed by gravity or by erosion of bedrock in place.

Deposition Arches

Rock is not always moved as consolidated blocks. Water can dissolve minerals from rock and transport them to another place where evaporation or other processes cause them to be deposited. The most common example is the familiar white ring left inside a glass as the hard water it holds evaporates, leaving behind an accumulation of calcium carbonate previously dissolved from calcareous rock. The same process on a larger scale produces stalactites and stalagmites in caves, and flowstone and travertine mounds around springs. Under the right conditions, it can also produce an arch. A large example once graced Jughandle Falls in Clifton Gorge (Greene County). It formed where a stream highly charged with lime from calcareous glacial till plunged over the lip of a waterfall. Calcium carbonate was deposited as a travertine mound on slump blocks at the base of the falls and as a curtain extending down from its lip. When the mound and the curtain met, an oval hole a few feet in diameter was left between the resulting column and the face of the cliff behind the waterfall, forming the Jughandle. Unfortunately, this interesting deposition arch collapsed around 1900. Since then, travertine has continued to form and the curtain and mound are being rebuilt. Should they ever reconnect, the Jughandle will be reborn. It is possible for such deposition arches to grow very large. One example in Arizona has an opening 140 feet wide, 60 feet high and 400 feet long and actually spans a river, forming a natural bridge. The example shown in Figure 35 is much smaller, having a span of 3 feet. It is actually a compound gravity and deposition arch. A rock lodged between a slump block and the nearby cliff formed the original arch. This was then cemented in place by a coating of calcium carbonate deposited by spring water.

Man-made "Natural" Arches

At least one example of a man-made "natural" arch was observed during the course of this study (Figure 36). During the operation

Figure 35. A deposition arch. This is actually a compound arch which originated when a small rock lodged against the cliff. The rock was then cemented in place by a coating of calcium carbonate.

Figure 36. Castalia Quarry Reserve Arch, a rare example of a man-made "natural" arch.

of a dolomite quarry near Castalia, blasting knocked out a section of weaker rock from between two vertical crevices. A layer of thin-bedded rock above the crevices remained in place, forming a lintel. The arch which resulted is very blocky in appearance and quite unstable. The lower beds of the lintel are sagging and will no doubt eventually break and collapse. Quarry operations ceased before this feature was destroyed and the area has now become part of Erie County MetroParks' Castalia Quarry Reserve which is accessed from State Route 101. The arch can be readily seen from the Quarry Rim Trail. Due to the instability of the quarry walls in general and of this arch in particular, it is best viewed from a distance.

Chapter 3

A Catalog of Ohio's Natural Arches

The following list of natural arches formed in Ohio was developed over a period of some twenty-five years. Even so, it is not complete. Several potential arches have yet to be located and surveyed and the existence of others not yet known is almost certain. The arches listed and described here are organized by geological age and geographic location. A complete list of these features is presented in Appendix I.

Arches in Ordovician Strata

Ordovician rock, the oldest bedrock exposed in Ohio, is found in the southwestern corner of the state (Figure 7). Composed of thin-bedded limestones and shales laid down in a warm shallow ocean, these rocks are known around the world for the abundance and quality of the fossil sea life they contain. No natural arches suitable for listing have been found in Ordovician strata in Ohio to date, due no doubt to their thin-bedded nature.

Arches in Silurian Strata

Silurian rock covers much of the western third of Ohio except for the southwestern and northwestern corners (Figure 7). As the bodies of lime-secreting animals living in the Silurian sea accumulated on the ocean floor, they formed thick deposits that hardened into limestone which over time was changed into dolomite, a calcium-magnesium carbonate. In the process, most of the original shells and other remains were dissolved away, creating the characteristic porosity of

most of Ohio's dolomite layers. Mound-shaped coral reefs are frequently found and their concentric, rounded layers appear to have
played a part in the formation of some of Ohio's natural arches. Three
formations, the Peebles Dolomite, the Greenfield Dolomite and the
Cedarville Dolomite, contain most of the natural arches and pillars
found in Ohio's Silurian rock.

Peebles Dolomite

The Peebles Dolomite is one of the major arch-forming units in
Ohio. This may be due in part to the intensity with which it has been
studied, but the qualities of the rock itself are the most important
factor. Peebles Dolomite is generally massive, porous and vuggy (full
of holes, like Swiss cheese). This combination of characteristics enables it to stand in vertical cliffs while still allowing the easy passage
of large amounts of groundwater, two conditions important in the formation of natural arches.

Nowhere is the cliff-forming tendency of the Peebles Dolomite
better shown than in the region where Adams, Highland, Pike and
Ross counties meet. Here it reaches its greatest thickness, approaching 90 feet. Here, too, the great Pleistocene ice sheets reached their
southern limits. The powerful melt water streams which poured off
them as they wasted away were important agents of erosion, carving
deep canyons into the resistant bedrock. The gorges of Rocky Fork,
Baker Fork and Ohio Brush Creek as well as the bluffs lining many of
the smaller tributaries of the region owe their present impressiveness
to this combination of hard dolomite and torrents of melt water.

It is in these bluffs and gorge walls that the arches of the Peebles
Dolomite are found. They cluster in the deepest gorges where the
highest cliff faces and largest exposures of rock occur. The greatest
local concentration of arches in the entire state is found in the gorge
of Baker Fork, much of which is fortunately protected within the
boundaries of Fort Hill State Memorial operated by the Ohio Historical
Society. Scioto Brush Creek and Rocky Fork Gorge also shelter concentrations of natural arches that have formed in the Peebles Dolomite.

The striking ability of Peebles Dolomite to form arches can be
credited to its past history. This formation contains numerous reefs

which indicate that it formed in a shallow sea. Retreat of the sea exposed the dolomite to weathering under semi-arid conditions which dissolved the soluble rock, creating typical surface and subsurface karst features. These include caves, sinkholes, underground drainage channels, vugs (cavities in the rock, often with a mineral coating that differs from the surrounding bedrock), solution-enlarged joints, collapsed strata and breccias (rock made of coarse, angular fragments, in this case probably the result of cave collapse). Eventually, rising seas or subsidence of the land submerged the eroded rock once again, allowing deposition of the Greenfield Dolomite over the weathered Peebles.

Renewed weathering of the Peebles during the present cycle of uplift and erosion has accentuated weaknesses in the rock fabric where it is exposed. Breccias have been removed, vugs and other openings have been enlarged and caves have been breeched, giving the weathered Peebles its familiar "rotten rock" appearance. Each of these processes can result in the formation of natural arches. It is interesting to note that the greatest concentration of paleokarst ("ancient karst") features in the Peebles is found in its upper part, just where most of the natural arches are located.

Arches of Baker Fork Gorge

The walls of Baker Fork Gorge in southeastern Highland County hold eight listed arches, including one of Ohio's natural tunnels — the largest cluster of arches in one small area found to date in Ohio. At first glance, Baker Fork seems to be a misguided stream, running as it does from the level plain of Beech Flats toward the bold hills at the plain's southern edge. More surprising yet, it appears to force its way through the barrier, flowing in an impressive chasm sliced across the hills. How it managed to accomplish this feat is a story that can be repeated for dozens of other streams in Ohio.

What is now Baker Fork was originally two streams, both of which headed in coves on opposite sides of a ridge which once connected Fort Hill with the hills to the west (Rosengreen, 1974). The unnamed stream flowing north from this divide made its way to a preglacial version of Paint Creek by way of a wide, deep valley now

83

occupied by Beech Flats. On the other side of the divide, ancestral Baker Fork flowed south, following in general the course of present Baker Fork.

As the advancing Illinoian glacier covered ancestral Paint Creek, it formed an ice dam that ponded meltwater in what remained of the valleys of the creek and its tributaries. The rising waters of the lake crept inexorably up toward the headwaters of these streams until it reached the level of the lowest gap in the hills confining it. That gap, or "col," happened to be in the ridge between the heads of ancestral Baker Fork and the now-flooded unnamed tributary of Paint Creek.

As glacial meltwater from the rising lake poured through this col and into the valley of ancestral Baker Fork, it cut down into the soft shale capping the former divide. The deepening channel allowed the water to drain faster which in turn enabled the newly formed river to erode with even greater force. With the whole weight of the pent-up lake behind it, the newborn river quickly cut through the shale and deep into the Peebles Dolomite below. When the last of the lake finally drained away, a raw, new chasm — Baker Fork Gorge — sundered Fort Hill from its neighbors to the west.

During the glacial lake's brief existence, silt and rocky debris released by the melting glacier and eroded from surrounding hills washed into it and settled to the bottom, filling the old valley of ancestral Paint Creek and its tributaries to depths as great as 240 feet. The surface of this fill now forms the wide expanse of Beech Flats. After the glacier melted out of the area, local drainage flowing onto the raised floor of the extinct lake was diverted south through the gorge. Thus was born the stream we now know as Baker Fork. This tale of the glacial diversion of preglacial drainage was repeated at many places along the ice front; many of the more spectacular gorges and river "narrows" of the state were formed in this way.

Beneath the shale of the bisected ridge lay the solid platform of Peebles Dolomite on which the hills stand. The raging glacial river cut down as much as 60 feet into this tough bedrock, creating sheer cliffs. Continued erosion along local lines of weakness has broken much of the cliff line into rounded knobs separated by short, steep gullies. It has also taken advantage of fractures, bedding planes and areas of weakened resistance in the rock to form the natural arches which are among the gorge's more interesting features.

Boundary Arch (OH-A-HIG-01) — Highland County
Span: 5.67 feet (1.3 meters); Clearance: 9 feet (2.7 meters)
Boundary Arch (Figure 37), named for its location near the boundary of Fort Hill State Memorial, is located on the west side of Baker Fork Gorge in the angle of a large, rectangular reentrant (an offset of the cliff line which extends inward). The cliff here rises only 13 feet above the bottomland of the stream and the arch with its lintel occupies the full height. The reentrant was formed by the removal of a block of dolomite along two vertical joints, one paralleling Baker Fork and the other perpendicular to it. The arch represents a widening of

Figure 37. Boundary Arch. A view of the front entrance showing the vertical crevice which has been widened to form the arch.

the perpendicular joint beyond the removed block. At the face of the arch, this widening affected only the bottom 9 feet of the joint, leaving a lintel that is 4 feet thick and 2.6 feet wide at the top of the cliff to form the arch. Beyond the lintel, solution has enlarged the fracture through the entire thickness of the cliff resulting in a narrow, linear skylight 1 to 2 feet wide and 25 feet long perpendicular to the lintel. The back wall of the skylight slopes steeply down from the level of the cliff top to the floor of the arch which is continuous with the floodplain of the creek. The opening beneath the lintel is triangular in shape, narrowing at the top into a barely perceptible crevice.

A small seep at the upslope end of the skylight may have been a factor in enlarging the crevice forming the arch. On the other hand, the arch could have been formed rapidly by solution and erosion when the gorge was being cut, especially if groundwater movement through the crevice now represented by the exposed seep had already weakened the surrounding rock. The cliffs on the opposite side of Baker Fork are almost twice as high as those at Boundary Arch. This would seem to indicate that the area of the arch was under water during drainage of the glacial lake that filled Beech Flats. The violent outflow could have been responsible for breaking up and carrying away the block of dolomite which once filled the reentrant. At the same time, the escaping water could have been seeping down into the crevice behind the present arch and enlarging it. If such were the case, Boundary Arch could be almost as old as the gorge.

Whatever process enlarged the crevice, it left the lintel of the arch intact. Less-resistant rock or more open channels for water movement must have existed below the level of the lintel. The skylight does not appear to have ever been roofed and so the arch cannot be a result of the incomplete collapse of a chamber.

Natural Y Arch (OH-A-HIG-02)* — Highland County
Span: 29 feet (8.8 meters); Clearance: 5 feet (1.5 meters)
Natural Y Arch (Figure 38) is unusual in that it has a double skylight which gives its roof a Y shape. It was originally named Natural Y Bridge for its fancied resemblance to the famous Y Bridge built over the junction of the Muskingum and Licking rivers in Zanesville,

Figure 38. Natural Y Arch. The interior as viewed from the entrance showing the double skylight.

Ohio. However, this arch does not cross a valley of erosion and is therefore not a natural bridge; in order to avoid confusion, its name has been modified.

The arch is located on the west side of the gorge about 45 feet above Baker Fork. It is directly opposite the mouth of Beech Ravine and is readily visible from Gorge Trail after the trees lose their leaves in the autumn. It has a span of 29 feet and a clearance that varies from 10.5 feet at the front to 5 feet at the back. The lintel is 20 feet wide and from 3.5 feet to 5.75 feet thick.

A linear skylight 8 to 10 feet wide parallels the rear of the lintel. Its form and location appear to be joint controlled. A shallow linear depression in the hillside extends from the upstream end of the skylight around the neighboring abutment and into a narrow gully leading down to Baker Fork. This may be the remnant of a channel formed by wash off the steep hillside above. This wash, pirated by the joint and then following a bedding plane out to the face of the cliff, may have been the principle factor in forming the arch.

The identifying characteristic of Natural Y Arch is the presence of a second, much smaller skylight piercing the roof of the alcove at its downstream end. This skylight is 14.7 feet long and 4.7 feet wide. The roof remnant between the two skylights varies in width from 6 feet where it joins the hillside to 2.3 feet where it connects to the lintel. At 3 feet thick, it is noticeably thinner than the adjacent edge of the lintel. Like the main skylight, the secondary skylight is joint controlled as indicated by its narrow, linear form and vertical walls. Erosion has not yet proceeded far enough along this minor joint to sever the small limb from the lintel. Although no continuous drainage appears to be flowing through either opening at present, sheet wash does utilize them. Each opening lies at the bottom of a gouge in the hillside, the result of years of intermittent drainage through these holes into the gorge.

Natural Y Arch may be related to Davis Cave, a small solution feature found in the side of a minor tributary valley which enters Baker Fork just upstream from the arch. The main passage of the cave runs in a direction which would, if continued, bring it out into the main gorge at or near Natural Y Arch. The cave and the arch may represent segments of a system of small solution channels cut in a vulnerable layer of rock. When these channels were severed by Baker Fork and its tributary, atmospheric erosion could have enlarged their openings to form the cave and the arch. Against this possibility are the short length (only 25 feet) of Davis Cave and the lack of evidence for any such channels in the back wall of the skylight of Natural Y Arch.

A hermit named David Davis reportedly once lived in a cave in the gorge and mined clay from which he made paint to sell (Overman, 1900). Although Davis Cave was named for this reclusive entrepreneur, a map printed with Overman's account shows his cave (there called Hermit Cave) as being nearer to where Natural Y Arch is located. Since in popular usage the term "cave" can also refer to recess shelters or alcoves such as Old Mans Cave in Hocking County, Natural Y Arch was probably the cave referred to by Overman. The arch, located as it is on a steep hillside, would have made a much drier, more pleasant and — of great importance to a hermit — less accessible home than Davis Cave.

The Keyhole (OH-A-HIG-03)* — Highland County
Tunnel Length: 77 feet (23.5 meters); Tunnel Width: 2 feet (0.6 meters)

The Keyhole (figures 39 and 40) is exceptional, not only for being one of the largest arches in Ohio, but also for being one of the rarest of arch types, a natural tunnel. Its name was given long ago. During a 1923 visit to Baker Fork Gorge, E. Lucy Braun, one of Ohio's premier early ecologists, made the "Keyhole Bridge" her destination. It was most likely so called from the peculiar shape of its opening as seen from either end. Being wide and somewhat rounded at the top,

Figure 39. The Keyhole. The front entrance which faces Baker Fork.

and tapering irregularly toward the bottom, it bears a vague resemblance to an old fashioned lock of the type opened by a skeleton key.

The Keyhole is obviously an enlarged vertical crevice. It cuts through a narrow ridge separating Baker Fork from the small, steep-walled valley of Bridge Creek which parallels the main gorge. This valley is totally enclosed. The only ways into it are through the tunnel or down its steep sides.

The height of the tunnel was estimated to be 35 feet. Were it a true cave passage, it would be among the largest known in the state. It is long enough to protect its inner reaches from most atmospheric disturbances and so allows the growth of typical cave formations such as flowstone, small stalagmites and botryoids which are small, rounded deposits of calcium carbonate resembling bunches of grapes.

While the center of the tunnel is protected from atmospheric weathering, the ends are not. Both entrances have been enlarged enough to break through the ceiling, creating open-air "vestibules." By this means, 34 feet of the upper end of the tunnel and 17 feet of the lower end have been opened. Neither of these vestibules is included in the recorded length of the tunnel.

A seep line running the length of the tunnel on both sides has helped to widen it. Here the constant wetting and drying of the rock by escaping groundwater has caused the walls to retreat 2 to 5 feet, providing the largest of the notches which gives the opening silhouette its keyhole shape. Much of this widening may be due to actual solution of the rock, as shown by the deposition of travertine armor which coats the wall below the notch. Where the seep is still active, travertine is still being deposited.

Although the part of the watershed of Bridge Creek that feeds into the tunnel is not large, it collects drainage from a major part of the southeastern flank of Jarnigan Knob. Water was found flowing in the lower end of the valley and through the tunnel during every visit. Within the tunnel itself the stream is only 2 inches deep. Even so, it enables this arch which is also a tunnel to be considered a true natural bridge.

The bridge itself (the lintel of the arch) is a steeply rounded ridge falling off to Baker Fork on one side and into the enclosed valley of Bridge Creek on the other. Deer Trail passes over it, making it a bridge in fact as well as form. The upper third of the ridge is made of easily eroded Ohio Shale, creating the rounded top of the lintel. The bottom two-thirds is Peebles Dolomite which appears as high, cliff-like bluffs on either side of both entrances to the tunnel.

The most intriguing question about the tunnel concerns its origin. Elsewhere in Baker Fork Gorge, fractures in the Peebles Dolomite which run perpendicular to the stream result in short, steep, narrow ravines separated by headland-like cliff remnants. At The Keyhole the same conditions yielded something strikingly different. Why?

A look at the topographic map of the area suggests one possible explanation (Figure 40). On it a strange, L-shaped valley is shown 700 feet downstream from the tunnel. Its shape contrasts sharply with the straight lines of the other small valleys leading into Baker Fork Gorge, but mimics on a more angular scale that of the enclosed valley leading into The Keyhole. This may indicate that these two valleys share a common history.

If the lower L-shaped valley's upright stem which parallels Baker Fork were extended upstream, it would be continuous with the enclosed valley of Bridge Creek. Close inspection of the rim of the enclosed valley shows that this connection actually exists. A barely perceptible notch in the southern rim of the valley leads into the upper end of a very shallow but fairly broad trough which deepens to the southeast, becoming the upright stem of the lower valley. The upper reaches of this shallow valley are dry, but two-thirds of the way down to the right-angle bend of the L it begins to fill with local drainage which appears first as soggy ground and then as a tiny stream flowing down the L-shaped valley into Baker Fork.

Given these observations, one possible explanation for the existence of The Keyhole would be as follows (Figure 41).

(A) In pre-Illinoian time ancestral Bridge Creek was a tributary of ancestral Baker Fork. Both were slowly eroding their way into the south side of the shale hills forming the old divide — Bridge Creek into the flank of Jarnigan Knob and ancestral Baker Fork into the col between Jarnigan Knob and Fort Hill. At this time, the course of ancestral Bridge Creek ran from its present head down across the site of

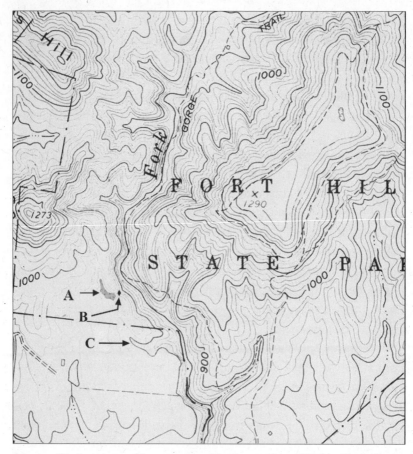

Figure 40. A topographic map showing the section of Baker Fork Gorge that contains The Keyhole in Fort Hill State Memorial (From US Geological Survey Sinking Spring, Ohio, Quadrangle Map; contour interval = 20 feet). **A:** The enclosed valley behind, or upstream from, **B:** The Keyhole. **C:** The L-shaped valley downstream from The Keyhole. The Fort Hill parking lot is located approximately 0.1 mile beyond the northeast (upper right) corner of the map. Trails on the memorial are shown by short dashes.

the present enclosed valley and through the shallow trough to the L-shaped valley and so into Baker Fork.

(B) With the advance of the Illinoian glacier and the formation of Beech Flats Lake, the cutting of Baker Fork Gorge across the col began. As the outlet stream cut lower, the local base level of erosion

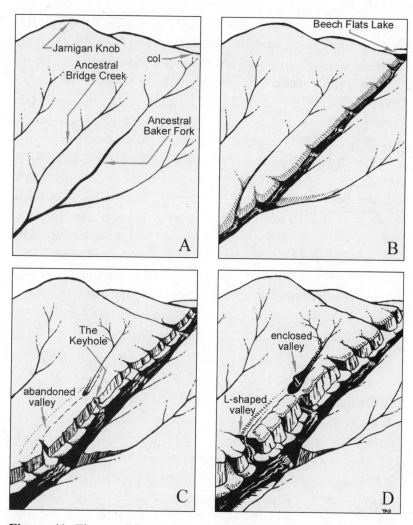

Figure 41. The possible history of formation of The Keyhole as explained in the text.

was also lowered. Ancestral Bridge Creek and the other tributary streams were rejuvenated and began cutting their own valleys deeper.

(C) As it cut through the Ohio Shale and into the Peebles Dolomite, Ancestral Bridge Creek found a more direct outlet into the deepening gorge of present Baker Fork through the upper level of a joint crossing its path. This subterranean flow offered Bridge Creek a shorter path and greater fall into Baker Fork, giving it more erosive power and enabling it to widen the crevice downward at about the same rate that Baker Fork Gorge was being lowered, forming The Keyhole. Meanwhile, the now isolated drainage of the lower section of Ancestral Bridge Creek was also captured by a joint perpendicular to Baker Fork Gorge. In this case, rapid down-cutting of the gorge end of the crevice resulted in a typical open ravine rather than an arch, perhaps due to the lack of an Ohio Shale cap over the Peebles Dolomite at this point.

(D) With final retreat of the Illinoian glacier and draining of Beech Flats Lake, Baker Fork assumed its modern form. Bridge Creek lowered the floor of the tunnel to its present depth several feet above the level of Baker Fork, resulting in the picturesque cascade at the entrance to The Keyhole. Both ends of the tunnel were enlarged by atmospheric erosion to form the vestibules. Downstream, the head of the lengthening open ravine reached the valley abandoned by Bridge Creek and completed the capture of local drainage that was using it, giving the pieced-together valley its unusual L-shaped bend.

Baker Fork Arch (OH-A-HIG-04) — Highland County
Span: 15.7 feet (4.8 meters); Clearance: 8 feet (2.4 meters)
From the hillside to the east, Baker Fork Arch (Figure 42) looks like a long, thin dolomite lintel perched on two abutments standing above its surroundings. It stands on the edge of an undercut bluff which drops 35 feet from the base of the arch to the talus slope below. The arch has a span of 15.7 feet and a clearance varying from 8 feet at the back of the arch to 10 feet at the front. The lintel is 12 feet wide and only 2 feet thick at the center of the arch. Its skylight takes the form of an oval depression in the hillside behind the arch. The back wall of the skylight joins the hill in a continuous slope.

This arch owes its existence to the greater resistance of the rock forming it as compared to the rock which once surrounded it. At some point in time, drainage found its way into the gorge through bedding channels beneath the span. A groove recessed 1.5 feet into the bottom of the lintel and extending across its entire width may represent the earliest stage of this drainage. Since no fracture is apparent in the top of the groove, it may have started as a bedding-plane opening.

As Baker Fork Gorge deepened, this drainage cut downward to create the original arch opening which has since been widened by

94

Figure 42. Baker Fork Arch. Baker Fork is visible through the opening of the arch.

atmospheric erosion. At the same time, the rock behind it was also removed, leaving the arch isolated. This removal may have taken the form of a cave which hollowed out the bluff behind the arch and then collapsed, leaving the present large oval skylight. Another bluff just downstream exhibits this very type of cave formation. However, no rock except that forming the Baker Fork Arch itself appears in the gently sloping walls of the skylight and there are no broken roof blocks on its floor. This would seem to indicate that the skylight was formed by solution from the top down and was never roofed.

Another possible explanation of the skylight involves the tendency of the upper layers of Peebles Dolomite in the Fort Hill region to crumble into a soft regolith (loose, unconsolidated rock material) locally called "marl," a term more commonly used for calcareous clay. This layer consists of a spongy, light grey to tan rock which weathers into tiny blocks 1/4 inch to 1 inch across. In this condition, it erodes into gently rounded slopes. Parts of this layer, however, are made of stronger rock and are more resistant to erosion, forming knobs rising from the surface of the harder, lower layers of the Peebles Dolomite. Such is the origin of Fort Hill Tea Table (Figure 166). Baker Fork Arch stands near the junction of the soft upper and hard lower layers of Peebles Dolomite and may represent a knot of harder rock which has resisted erosion better than the so-called "marl" layer which once surrounded it. This would help explain the slope leading down to the arch and the scooped-out shape of the skylight behind it.

A third possible explanation of the skylight relies on the past erosional history of the Peebles Dolomite. During its exposure prior to deposition of the overlying Greenfield Dolomite, the Peebles developed fairly extensive paleokarst which included enclosed sinkholes on its surface. Baker Fork Arch's skylight could represent one of these sinks reopened and modified by modern erosion.

E. Lucy Braun (1969) pictured Baker Fork Arch in her monograph on the vegetation of Fort Hill State Memorial. In it, she mentions that the sides of the arch were festooned with Canada Yew (*Taxus canadensis*) and *Andropogon*, a grass common to prairies. Although Canada Yew still festoons the arch, the *Andropogon* has disappeared, most likely shaded out by the growth of nearby trees.

Spring Creek Arch (OH-A-HIG-05)* — Highland County
Span: 18 feet (5.5 meters); Clearance: 5 feet (1.5 meters)

Spring Creek Arch (Figure 43) is the most easily observed Fort Hill arch. Only narrow Spring Creek stands between it and Gorge Trail. It is also one of only two Baker Fork arches located along a tributary rather than in the main gorge.

This arch has a span of 18 feet and a clearance which varies from 5 to 6 feet. The lintel is 11 feet wide and 5 feet thick. Its skylight, like that of Baker Fork Arch, is an oval-shaped depression whose back wall joins imperceptibly with the slope of the hill behind it. Low saddles at the upstream and downstream ends of the skylight join the hillside to the abutments of the arch. Sheet wash is actively enlarging the skylight, as shown by an apron of soil crossing the arch floor and descending as a wide talus slope to the stream below.

Spring Creek Arch has the appearance of being a large remnant of "rotten" rock. The downstream abutment is riddled with openings which could themselves be considered small arches. The largest of these is 2 feet long and 6 inches high. The Peebles Dolomite at this location, in contrast to its usual form elsewhere in the memorial, is made of several layers of rock with differing resistances to erosion.

Figure 43. Spring Creek Arch. The arch as seen from the oval-shaped depression. Spring Creek can be seen through the opening.

The layer containing the opening of the arch and the smaller arches of the downstream abutment is more vulnerable than the layers above and below which form the lintel and floor of the arch, and has been cut back to form recesses beneath the resistant upper layer in both abutments. The resistant layer of dolomite forming the floor of the arch continues as a ledge beneath these recesses.

Drainage must have made its way down through the hard lintel layer of dolomite into the weaker rock beneath, enlarging whatever openings it found. This process might have begun as underground seepage even before the Illinoian glaciation; if so, it must have progressed at a painfully slow pace given the present small size of the arch. The cutting of Spring Creek Gorge in conjunction with Baker Fork Gorge sliced across these underground drainage channels, opening them up to atmospheric erosion which then enlarged the openings to give us the present arch. Water stains and small botryoids on the inside walls of the arch show that some seepage still flows through the vulnerable middle layer of dolomite.

The Passage (OH-A-HIG-06) — Highland County
Span: 7 feet (2.1 meters); Clearance: 2 feet (0.6 meter)

The Passage (Figure 44) promises big things, appearing as a dark hole 12 feet wide and 8 feet high opening near the top of one of the protruding bluffs forming the eastern cliff of Baker Fork Gorge. Reaching this opening requires technical climbing equipment, but another entrance is located down a steep ravine at the side of the bluff. This one is smaller, being 11 feet wide and only 2 feet high. From this entrance an opening shaped like a reclining 7 extends into the bluff. The long leg of the 7, which begins at the smaller entrance, is 20 feet long and varies in height from 2 feet at the opening to 4 feet where it meets the wall at the angle of the 7. The increase in height is a result of the lowering of the floor and would be even greater were it not that the ceiling also drops. The shorter part of the void, the top of the 7, is 17 feet long and ends at the large entrance in the cliff. This entrance has been widened into a flattened oval, probably by an active seep line near its middle. Flowstone and very small active stalactites within the opening show that groundwater still gains access to it.

Figure 44. The Passage. The view out through the large entrance.

The level of the Peebles in which The Passage is found is pocked with solution holes. The walls of this arch are heavily pitted, especially near the larger entrance. One of these pits extends at least 10 feet completely through the wall and opens onto the cliff face above Baker Fork. This evidence of extensive atmospheric weathering shows that the rock around the larger entrance has been exposed for a longer time than that around the smaller entrance. The size of the large entrance itself is a result of long-term atmospheric weathering.

The small entrance, by contrast, is a long, narrow fissure. It has the appearance of having been opened recently by down-cutting of the ravine in front of it. On the opposite side of this ravine is another entrance similar in size and shape to the smaller one leading into The Passage. This one opens into Double Dome Cave, a small, dome-shaped void. Although The Passage is at a lower level than Double Dome Cave, the two may originally have been linked before the gully between them formed. If so, then deepening of the gully severed the original cave with the inner section becoming Double Dome Cave and the outer section becoming The Passage.

Big Cave Natural Bridge (OH-A-HIG-07) — Highland County
Span: 8 feet (2.4 meters); Clearance: 4 feet (1.2 meters)
Big Cave Natural Bridge (Figure 45), located high on a sheer cliff, provides the clearest demonstration of a fact which is easily overlooked at the other arches found in Fort Hill State Memorial; all of them have been formed in the uppermost levels of the Peebles Dolomite. The arch is located on the west side of the gorge at the very top of the cliff and was named for a large recess cave indenting the cliff below and slightly upstream from it.

The valley feeding into the skylight of Big Cave Natural Bridge is extremely short and shallow and presently carries water only at rare intervals. It drops into the southwest corner of the narrow, linear skylight which is 8 feet wide and 40 feet long and elongated parallel to the cliff face. The shape of the skylight and its vertical walls show it to be an enlarged fracture. It is considerably wider than the tiny valley entering it. The arch opening is located at the opposite (northern) end of the skylight from this little valley and pierces the wall of rock remaining between the skylight and the cliff face. It has a

Figure 45. Big Cave Natural Bridge. The view through the entrance from the skylight.

clearance of 4 feet at the skylight, but enlarges to 6.5 feet on the cliff side.

The heavy, slab-like lintel is 7 feet thick and 12 feet wide. It is solidly connected to the cliff at either end and is obviously part of it, continuing its topmost level across the arch. From the river, the opening beneath the span is the most striking feature of this arch. It appears as a large hole in the cliff face similar to hundreds of other holes seen in the gorge. However, this one attracts attention because of its large size and the fact that sky is visible through it from below. This opening also exhibits a horizontal elongation which shows that the apparently massive Peebles Dolomite is actually made of layers of varying resistance to erosion, this particular layer being more vulnerable than others.

This natural bridge is part of a peculiar drainage adjustment. Whenever water flows through this system, it runs southeast down the shallow feeding valley, drops into the skylight and turns abruptly northeast to run its length, then makes another right-angle turn back to the southeast to flow beneath the lintel and down the cliff. A shallow notch-like valley cut in the talus slope immediately below the arch shows that water does occasionally take this convoluted route. These sharp angles and dislocated drainages must be due to the skylight. As noted earlier, the valley feeding into it is considerably smaller. Rather than being responsible for forming the skylight, the drainage using this valley is probably a recent addition. The skylight may even predate the gorge of Baker Fork. We are at this point a good distance south of the shale and sandstone hills rising above the Peebles and well out onto the plain formed by the top of the dolomite. If ancestral Baker Fork followed the same route that its gorge does today, the skylight of Big Cave Natural Bridge would have been in a position to take some of its overflow, especially when it was in flood stage. The skylight might represent a fracture which was being enlarged by seepage from the stream when it was flowing at or just below the top of the Peebles Dolomite. As the ancestral Baker Fork Valley deepened, drainage entering the skylight, by then probably only sheet wash off the nearby plain, found access to the river through an area of more soluble or more fractured rock at its northern end and then out to the valley through the less-resistant horizontal layer of rock below. Continued drainage gradually opened the arch between the skylight and the cliff face.

By the time the overflow river from the glacial lake filling Beech Flats cut through the col at Fort Hill and thundered down the valley of ancestral Baker Fork, Big Cave Natural Bridge may have been high enough to escape its violence. It is also possible that during its highest flood stages, the glacial river submerged the arch and continued its enlargement. If this explanation is correct, then Big Cave Natural Bridge could be the oldest arch in the memorial, having a history extending back to before the Illinoian glaciation.

Hidden Arch (OH-A-HIG-08) — Highland County
Span: 3.8 feet (1.1 meters); Clearance: 2.2 feet (0.7 meter)
Hidden Arch (Figure 46) was named for its location in the side of a small, rock-bound valley opening onto the east side of Baker Fork Gorge. It is one of only two arches that do not open onto the main gorge, the other being Spring Creek Arch. It is a small arch with a lintel only 1 foot wide and 9 inches thick. The skylight is 3.83 feet

Figure 46. Hidden Arch as seen from above.

long and 2.75 feet wide. The opening appears to be an enlarged version of the pits which are so common in this level of the Peebles.

A short distance downstream another feature similar in size and shape to Hidden Arch opens in the low stream bank (Figure 204). However, the span of this "almost arch" is broken by a 3-inch-wide gap. It gives an indication of what will happen to Hidden Arch as erosion continues to narrow its lintel.

Arches of Rocky Fork Gorge

Like Baker Fork Gorge, Rocky Fork Gorge can be attributed to down-cutting by glacial meltwater streams during retreat of the Illinoian continental glacier. After retreating from the line which created the lake in Beech Flats, the ice made another stand along a northeast-southwest line just south of present Rocky Fork Gorge. This opened up a drainage route for meltwater to the northeast into the Paint Creek system, and Baker Fork Gorge was abandoned to local drainage.

The next retreat brought the front line of the Illinoian ice to a position just north of present Rocky Fork Lake. This uncovered a depression formed by the valley of preglacial Rocky Fork, now buried beneath glacial till, which rapidly filled with meltwater. The resulting lake was dammed to the east by an esker, a ridge of gravel deposited by a meltwater stream running through a tunnel in the ice during a previous stand of the glacial front. The lake found an outlet around the south edge of the esker, cutting its way down into the Silurian bedrock to form what is now Rocky Fork Gorge. In the process, it cut across and perhaps helped to form The Seven Caves which are such an important attraction of the area today. As the ice melted back farther, the glacial lake drained completely and the gorge of Rocky Fork, like that of Baker Fork, became a conduit for local drainage. Once the gorge was cut, the natural arches within it began to form.

Miller Natural Bridge (OH-A-HIG-09)* — Highland County
Span: 46 feet (14 meters); Clearance: 3.5 feet (1.1 meters)
The manner in which Miller Natural Bridge (Figure 47) was formed is fairly clear. An intermittent stream falling over the side of

Figure 47. Miller Natural Bridge. An example of a breeched-alcove natural arch in which the alcove was broken by the widening of a joint parallel to the face of the cliff.

the gorge eroded an alcove in the base of the cliff. At the same time, water from the stream seeped into a crevice running parallel to the cliff face and 10 feet behind it, eventually making its way out the cliff face and helping to enlarge the alcove. Constant wetting and drying coupled with solution eventually enlarged the crevice enough that it captured the stream entirely, allowing it to drop through the roof of the shelter. The crevice has now been enlarged into a skylight 46 feet long and 11 feet across at its widest point, which also happens to be where the stream enters it. The remaining roof of the alcove forms the lintel of the natural bridge, appearing as a heavy slab of dolomite 4 to 6 feet thick and 10 feet wide.

The crevice forming the skylight can be traced into the west abutment of the natural bridge where it has been enlarged into a narrow "cave" a few feet wide at the entrance and extending 9 feet into the rock before finally pinching out. Active flowstone and botryoids on the walls of the cave show that water still moves through the bedrock. Beyond the west abutment, the back wall of the crevice becomes the cliff face. The west end of the abutment forms a sharply angular offset extending 20 feet toward the river. The front of the bridge marks the start of a new cliff line which continues eastward until broken by the entrance of a major side stream into the gorge. The dolomite here has a muddy look and contains casts of both horn and honeycomb corals.

The tiny, intermittent stream responsible for this impressive work has been active enough to build a small detritus fan, unfortunately enlarged by an old trash dump, where it drops into the skylight, and has cut a small valley below it to the river. A shallow groove across the top of the bridge at its narrowest point and in line with the present stream evidently marks the water's original path over the cliff.

Miller Natural Bridge and Miller Arch, described below, are both found in Miller Nature Sanctuary managed by Ohio DNAP. Visitors must obtain a free access permit before entering the area.

Miller Arch (OH-A-HIG-10)* — Highland County
Span: 9.5 feet (2.9 meters); Clearance: 5 feet (1.5 meters)

Miller Arch (Figure 48; Plate 1) may not be very large, but it makes one of the grandest entrances of any arch in Ohio. An unsuspecting

Figure 48. Miller Arch. The arch as seen from the steps descending to Rocky Fork.

visitor following Falls Trail in Miller Nature Sanctuary comes upon it suddenly — a great, rounded frame enclosing a cameo portrait of Rocky Fork Gorge. The Miller family, former owners of the preserve, called it Picnic Arch, reflecting their favorite use for it — an activity no longer permitted.

Part of the impressive appearance of this arch is due to its location at the junction of a small side gorge with Rocky Fork Gorge proper. The arch pierces a wedge-shaped extension of the cliff projecting into the entrance of the side gorge. Its heavy lintel is held up on one side by the massive wall of the cliff and on the other by a thin, sloping pillar resembling the flying buttresses found on Gothic cathedrals. This pillar is only 3 feet wide at its base, yet it holds up a lintel that is approximately 13 feet thick.

The opening of the arch is 9.5 feet wide at the bottom and 5 feet high. It forms a rounded, lopsided triangle having its highest point at the pillar and sloping down to meet the flat floor on the opposite side of the arch. This floor extends beyond the arch in both directions and is obviously the top of a more-resistant layer of dolomite.

At first glance, this arch might appear to be the remnant of a collapsed cave. This impression is further strengthened by the presence of two privately owned small caves extending into the opposite wall of the ravine upstream from the arch. The small stream responsible for the gully may have broken through the roof of a cave system as it eroded down to meet the level of Rocky Fork, leaving the arch and the two caves behind. Such a process has been used to explain the present appearance of The Seven Caves farther down the gorge.

Closer inspection of the dolomite framing the arch, however, shows it to be bedded in highly fractured layers which is not the type of rock in which caves usually form. Furthermore, these layers are curved, paralleling the shape of the opening. Such contorted and brecciated beds tilted in different directions have been found in other exposures of the Peebles Dolomite. These localized variations in the otherwise massive dolomite may be an expression of the paleokarst found in the upper levels of this formation. Another example found in the small stream leading away from Trimmer Arch several miles north of Miller Nature Sanctuary has the appearance of a sinkhole or small cave which collapsed prior to deposition of the overlying Greenfield Dolomite. The localized angular bedding found at Miller Arch may be due to a similar cause, or it could be the remains of a mounded coral reef which are often found in Ohio's Silurian bedrock.

While the cause of this peculiar structure is obscure, its presence obviously had much to do with the formation of the arch. The fractured dolomite forming the base of the wedge-shaped cliff projection was attacked by the elements from both directions and readily gave way. The resistance of the pillar may be due as much to the coating of travertine armor which covers it as to any inherent strength in the rock of which it is formed. The zone of fractured dolomite is very local, extending only a short distance beyond the arch. Just upstream, the dolomite assumes its more typical massive structure and any bedding present is horizontal.

Jawbone Arch North (OH-A-HIG-11) — Highland County
Span: 8.5 feet (2.6 meters); Clearance: 4.7 feet (1.4 meters)

Jawbone Arch South (OH-A-HIG-12) — Highland County
Span: 4.6 feet (1.4 meters); Clearance: 3.3 feet (1 meter)

The two Jawbone arches (Figure 49), named for the jawbone of an ungulate found in the long alcove they share, are not obvious. The dark opening leading to them is easily seen, appearing as a deeper version of the undercut which marks the base of the cliff in this part of the gorge. Only after stepping inside does it become apparent that a widened vertical crevice at its back has broken through to the surface in two places.

In form, the Jawbone arches are similar to Bundle Run Arch (Figure 60). All have formed in the Peebles Dolomite by an enlarged

Figure 49. Jawbone Arch North and Jawbone Arch South. The vertical crevice which has been widened in two places to form the arches parallels the cliff face, forming the back wall of this long, narrow alcove. The arch openings are not visible from this vantage point.

vertical crevice meeting a widened horizontal bedding plane. The alcove and skylights in both are very long and narrow. In the case of the Jawbone arches, the enlarged vertical crevice which parallels the cliff face is 66 feet long, but only 5 to 7 feet wide where it breaks through the surface of the upland above the arch. However, it narrows considerably as it descends and is only 5 to 10 inches wide where it forms the skylights. It then widens again until at the point where it meets the alcove, its width equals that at the surface, giving it the outline of a tall, skinny hourglass. The back wall of the alcove is nearly vertical, formed by the upland side of the crevice. It was originally thought that the two skylights were actually a single opening divided by a fill of loose debris. However, closer inspection showed that the vertical crevice has not weathered open through the entire length of the skylight, or even for a majority of it, and that the two openings are indeed two separate skylights creating two separate arches.

The roof of the alcove which is here considered to be the lintel of the arches is very heavy and blocky in appearance, being about 12 feet wide and thick. The deck is very flat and supports an amazing assemblage of spring wildflowers, as does the slope below the alcove. Were the skylight crevice widened enough to allow the lintel to project more freely from the cliff; it would look much like Miller Natural Bridge. Botryoids, flowstone and fingernail-sized rimstone dams — crescent-shaped calcium carbonate deposits outlining the lower edge of pools of water — on the sheltered back wall show that the seeping water noted during the survey is a common occurrence. A shallow channel leading down the steep slope in front of the arch shows that it takes in enough drainage to form a small stream at times. There are no valleys leading into the top of the enlarged vertical crevice, however. The upland behind the cliff's edge is level for several hundred yards before rising into a gentle hill marking part of an Illinoian recessional moraine. The drainage coming into the vertical crevice of the arch is sheet wash from the small area of upland behind it.

Besides the jawbone, possibly of a horse, for which the two arches were named and which most likely found its way into the alcove by means of some scavenger large enough to carry it, a sleepy opossum was seen lodged in one end of the vertical crevice during the first visit. It left soon after our arrival. Although to our minds the Jawbone arches are too wet a place to get some rest, other creatures are evidently not so particular.

Arches of the Scioto Brush Creek Drainage

Scioto Brush Creek takes its rise within the Appalachian Plateau not far from the crossroads town of Poplar Grove. Six miles later it leaves the sandstone and shale hills behind to make a short loop across the eastern edge of Ohio's Bluegrass Region. This is karst terrain, floored by the Peebles Dolomite. The gently rolling plain is pitted with short caves and shallow sinkholes. These evidences of carbonate solution are much smaller and less dramatic than those found in the more developed karst regions of Kentucky and Indiana, for the Peebles Dolomite is not as soluble as the limestone of those places. Still, Scioto Brush Creek has managed to entrench itself 20 to 50 feet into the dolomite and this, combined with several millennia of erosion, has allowed the formation of a subdued karst landscape. Among the features to be found within this part of the drainage basin of Scioto Brush Creek are several natural arches.

Riverbend Arch (OH-A-ADA-01) — Adams County
Span: 12 feet (3.6 meters); Clearance: 24 feet (7.3 meters)

Through much of its course across the Peebles Dolomite, Scioto Brush Creek swings back and forth within a narrow, steep-walled valley, many parts of which are lined with cliffs. Riverbend Arch (Figure 50), named by the Ohio Cave Survey during its reconnaissance of the valley, is found where one of these meanders brings the creek head-on into a 30-foot-high bluff, forcing it to make a right-angle turn. A vertical fracture splits the cliff at this point and extends back into the bedrock on the same course as the approaching river. The rock on one side of the fracture has been eroded back farther than that on the other, creating an angular offset in the cliff. Whether this feature is a result of recent erosion by the river or of past events is not known. Its present effect is to create a backwater where mud and debris tend to collect.

Figure 50. Riverbend Arch. The lower, elongated hole is the opening of the arch which slopes up to the skylight located on the upland above the upper hole in the cliff. Some of the school desks filling the skylight can be seen at the bottom of the lower hole.

Above this bar of muck rises Riverbend Arch, an enlargement of the fracture responsible for the offset. It appears as an opening in the cliff 12 feet wide at river level and rising about 24 feet to a Gothic-arched top. The crevice can be followed approximately 3 feet farther up the face of the cliff where it enlarges into a circular opening. The floor of the arch rises steeply from the mud of the river to a roughly circular skylight 32 feet wide and 28 feet long which opens less than 10 feet from the edge of the cliff. The mud, the steepness of the arch floor and, most discouraging of all, a mass of rusting school desks which have been dumped into the skylight prevented detailed study of the interior of the arch. It is not known whether the circular enlargement of the crevice above the river opening communicates with the arch or is a blind alcove.

Riverbend Arch could have had its genesis in groundwater drainage following the vertical crevice. After exposure by the cutting of the river valley, the crevice would have been enlarged by atmospheric erosion encouraged by moisture from the river. For some reason, this enlargement proceeded upward at an angle, creating the present skylight. It is also possible that the enlargement is a recent phenomenon initiated by the down-cutting of Scioto Brush Creek and keeping pace with it. If the sloping floor of the arch is solid bedrock, then a recent origin might be indicated. If, however, the arch opening continues as a crevice into the dolomite below the debris in the skylight, then a longer developmental history might be involved. The answer lies buried beneath all those school desks.

Roadside Arch (OH-A-ADA-02) — Adams County
Span: 4.5 feet (1.4 meters); Clearance: 3.5 feet (1.1 meters)
Roadside Arch (Figure 51), also named by the Ohio Cave Survey, is located at the side of a road near Bacon Flat. Here the valley wall of Scioto Brush Creek is a rough, pitted cliff about 15 feet high overlooking a narrow strip of overgrown bottomland. At one end of the cliff bordering the road is found Roadside Cave, 20 feet long, less than 3 feet high at the entrance and, on at least one occasion, occupied by bats. Roadside Arch lies at the other end of the cliff. Between them are Small Roadside Arch and Smaller Roadside Arch.

The entrances to both Roadside Arch and Roadside Cave have a circular outline and this similarity may indicate a common origin. The front opening of the arch has a 10 foot reach and a clearance of 7.5 feet which is rapidly reduced by a steeply sloping floor to just 3.5 feet at the back of the arch. The floor merges into the back wall of a skylight 8 feet long and 4.5 feet wide. The rim of the skylight actually lies 8 feet below the top of the rock. A small side stream cutting into the cliff beside the arch has carried away the upper levels of the Peebles Dolomite from around the skylight.

Like many other arches in the Peebles, Roadside Arch may have originated as groundwater channels or paleokarst features

Figure 51. Roadside Arch. Looking through the entrance toward the skylight.

which were exposed by stream entrenchment and then enlarged by atmospheric erosion. The back wall of the skylight is solid rock and gives no indication that this enlargement extends as a cave beyond the skylight.

Small Roadside Arch (OH-A-ADA-03) — Adams County
Span: 3.25 feet (1 meter) Clearance: 1.5 feet (0.5 meter)
    Unlike Roadside Arch which opens perpendicularly through the top edge of the cliff, Small Roadside Arch (Figure 52) opens parallel to the cliff face. It is found in a buttress extending out from the cliff at road level. The outside edge of the buttress forms an abutment which is only 2 feet wide.

Figure 52. Small Roadside Arch. The scale beside the arch opening is 12 inches long. Smaller Roadside Arch, marked by the arrow, pierces the rib of rock to the left of Small Roadside Arch. The dark opening at the far left is Roadside Cave.

Smaller Roadside Arch (OH-A-ADA-04) — Adams County
Span: 3 feet (0.9 meter) Clearance: 2 feet (0.6 meter)
Smaller Roadside Arch (Figure 53) is similar to Small Roadside Arch in every way, only smaller. It is about as small as an arch can be and still qualify for the Ohio list. However, in combination with the other two arches and the cave, it forms an interesting exhibition of some of the various expressions of erosion found in the Peebles Dolomite.

Scioto Brush Creek Arch (OH-A-ADA-05) — Adams County
Span: 25 feet (7.6 meters); Clearance: 8 feet (2.4 meters)
Scioto Brush Creek Arch (Figure 54) is located at the down-stream end of a dolomite bluff where the outside edge of a meander of Scioto Brush Creek has cut into the bedrock. Its thin, rainbow shape seems to spring from the cliff, making it one of Ohio's more graceful arches, especially when viewed from the river. The lintel is 4 feet thick and varies in width from 21 feet where it meets the cliff to 4 feet

Figure 53. Smaller Roadside Arch. Roadside Cave is visible beyond the arch. Small Roadside Arch is out of view to the right.

Figure 54. Scioto Brush Creek Arch. The view through the arch to Scioto Brush Creek from the collapsed portion of the alcove. The remaining alcove containing Covered Arch is beyond the right side of the photograph.

at the opposite end where it sinks into the ground. The presence of several large slump blocks at the narrow end of the arch show that this part of the lintel and the abutment it merges into were once much wider.

The skylight opening behind the arch is 16 feet wide and 50 feet long. The skylight is so much longer than the arch because the back side of the arch opening is continued as an alcove in the uphill side of the cliff abutment. This recess is quite deep and its back wall meets the floor in a ragged, scalloped line. The entire skylight may at one time have been roofed, forming a much larger and deeper alcove. Covered arch is found within this remnant alcove. The back line of the lintel is very straight, indicating fracture control. The rear portion of the roof could have given way along this plane of weakness, leaving the front of the roof as the arch. The upstream end of the rear wall of the skylight is marked by a low rock ledge indicating that it has recently been disconnected from the alcove beside it. The other half of the rear wall of the skylight is a slope with no rock visible and has suffered erosion for a longer time. This may have been the first portion of the roof to collapse.

An active seep coming out from the rear wall of the skylight and running through the arch to the river gives further credence to the theory that it was formed through alcove breaching. Such seeps are commonly found coming out of enlarged openings in the Peebles Dolomite. Usually these openings are vertical crevices. In this case, a layer of soluble rock appears to have allowed movement of water along bedding planes which, coupled with atmospheric erosion at the cliff face, created an alcove. Collapse of the rear portion of the roof resulted in the formation of Scioto Brush Creek Arch.

Covered Arch (OH-A-ADA-16) — Adams County
Span: 5.1 feet (1.5 meters); Clearance: 2.3 feet (0.7 meter)
The small Covered Arch (Figure 55) is found beneath the remnant alcove beside Scioto Brush Creek Arch. The back wall, ceiling and floor of the alcove frame three sides of the opening. A thin column barely 9 inches wide reaching from the floor to the ceiling forms the fourth side. Although the arch appears to be a result of atmospheric weathering, traces of flowstone on the column show that

Figure 55. Covered Arch. The entrance of the alcove which "covers" this arch is behind the viewer.

deposition has also occurred. The arch was named for its covered location within the alcove.

Cedar Fork Arch (OH-A-ADA-06)* — Adams County
Span: 11 feet (3.3 meters); Clearance: 13 feet (4 meters)

Cedar Fork, rising on the northwest slope of Peach Mountain, is an eastern tributary of Scioto Brush Creek. It falls 300 feet in 6 miles before joining the larger stream. Throughout its lower course it flows in a narrow valley cut into the Peebles Dolomite, part of which has been protected within Davis Memorial and Shoemaker state nature preserves.

Cedar Fork Arch (Figure 56) is located on the south bank of the stream on the lowest slope of Tolle Hill, within Shoemaker State Nature Preserve. Seen from the river, the arch has the appearance of being a remnant of the hill itself — a knot of tougher rock that has resisted the erosion which pushed back the hillside. In this it closely resembles Spring Creek Arch in Highland County. The opening of Cedar Fork Arch is extremely sinuous both vertically and horizontally.

Figure 56. Cedar Fork Arch. The view from Cedar Fork through the arch to the skylight.

Its ground plan resembles a backward S with a span of 11 feet at its narrowest point. The profile of the opening is also quite sinuous. The lower entrance mimics the ground plan in that it resembles a fat S with the addition of an enlarged crevice a few inches wide rising from the top.

On the uphill side, the opening expands into an alcove with a reach of 22 feet. The alcove opens onto the skylight and appears to be a covered portion of it. Whether or not the rest of the alcove was ever roofed is not known. The back wall of the skylight merges with the floor of the arch which falls steeply from back to front, providing a clearance beneath the lintel which varies from 7 feet at the back to 14 feet at the front. The lintel itself ranges in thickness from 2 to 3 feet. Even though the back wall of the skylight is fairly steep, it is not steep enough to prevent deer from making their way down it to utilize the arch as a passage through the line of bluffs to the river below, as shown by a muddy path and sharp-toed tracks. Even though a break in the cliffs a few hundred yards downstream offers an easier route to the stream and is more heavily used, a number of deer for whatever reason use the awkward trail through the arch.

Cedar Fork Arch is a result of the patient whittling away of surrounding rock. A stronger upper layer of dolomite has given a measure of protection to weaker layers beneath, although not enough to prevent their being cut back beneath the upper layer to form the alcove. The arch opening itself resulted from the piercing of this weak layer.

The area surrounding this arch contains several interesting features. Cedar Fork Standing Stone, a natural pillar, stands on a low bluff nearby (Figure 193). Mattress Arch is located in the upstream abutment of Cedar Fork Arch, and Crawl Arch and Blocked Arch are both found a short distance farther upstream. A number of state-listed rare plants grow in the immediate vicinity of the arch, including Resurrection Fern (*Pleopeltis polypodioides*), Heart-leaved Plantain (*Plantago cordata*), Sullivantia (*Sullivantia sullivantii*) and of course White Cedar, also called Arbor Vitae (*Thuja occidentalis*), from which the stream gets its name. Fortunately, the land containing most of Cedar Fork Arch as well as Mattress, Crawl and Blocked arches described below, has been donated to Ohio DNAP as a state nature preserve and so now has some measure of protection.

Mattress Arch (OH-A-ADA-07)* — Adams County
Span: 3 feet (0.9 meter) Clearance: 2 feet (0.6 meter)
The dolomite along Cedar Fork is pitted with solution cavities. A large one passes completely through the upstream abutment of Cedar Fork Arch to form Mattress Arch (Figure 57). From an entrance 2.7 feet wide and 2.9 feet high, the void enlarges into a dome-shaped room 3 feet high, 6 feet wide and 8 feet deep — large enough to hold an ancient, musty mattress which was seen during the survey's first visit. On a more recent visit, the mattress was found to have migrated out of the arch, but not far, and was resting on a sloping shelf 50 feet downstream, providing a most unstable surface for . . . something.

Crawl Arch (OH-A-ADA-08) — Adams County
Span: 4.2 feet (1.3 meters); Clearance: 3.5 feet (1.2 meters)
Upstream from Cedar Fork Arch, short gullies break the low, dolomite cliff along the south side of Cedar Fork into a series of bluffs

Figure 57. Mattress Arch. The arch opening as seen from Cedar Fork. The scale is 12 inches long. The mattress is gone.

similar to those found in the gorge of Baker Fork, only smaller. They rise only 15 feet above the stream and present a very ragged outline. They are, in effect, the solution-pitted, erosion-exposed bones of the hillside behind them.

Across the neck of one of these bluffs and parallel to the stream runs a joint which has been widened to form Crawl Arch (Figure 58). The opening extends 22 feet from the upstream end of the bluff to a narrow gully cutting across the joint. In cross-section, this arch is typical of enlarged vertical joints found in the Peebles Dolomite — wide at the bottom, but narrowing rapidly toward the top. In this case, most of the widening has occurred on the stream side of the long passage, giving it a decidedly lop-sided appearance. Atmospheric weathering has widened the upstream entrance of the arch to an impressive width of 9 feet and height of 5 feet. The interior of the arch constricts, however, giving a span of 4.2 feet and a clearance of 3.5 feet.

Figure 58. Crawl Arch. A view from the gully to the upstream entrance. The vertical crevice which was enlarged to form this arch can be seen in the upper left hand corner.

The enlarged joint continues on the far side of the narrow gully, which is only 4 feet wide. It has not been opened so much here as in Crawl Arch, but it does pierce the other section of the bluff and widens at the far end to become Blocked Arch.

Blocked Arch (OH-A-ADA-09) — Adams County
Span: 6 feet (1.8 meters); Clearance: 4 feet (1.2 meters)

Blocked Arch (Figure 59) has formed in the same joint as Crawl Arch, but at the opposite end of the bluff. Most likely the two arches were connected until the gully between them eroded down far enough to sever their union. Its larger entrance, which faces downstream, has the same lop-sided outline as does the main entrance of Crawl Arch. The widened crevice can be followed straight into the bluff for a distance of 26.5 feet where it meets the gully separating Blocked Arch from Crawl Arch. The gully cuts the crevice at a right angle and opens out onto the face of the bluff overlooking Cedar Fork. The mound of

Figure 59. Blocked Arch. The gully entrance is on the left, the downstream entrance is on the right.

rocks, dirt and tree parts that had washed down the gully and partially blocked this entrance at the time of the first visit was found to be greatly reduced during a second visit and the arch is no longer blocked. The name, however, was retained.

Just upstream from Blocked Arch and Crawl Arch towers the rubble wall of a major rock quarry. For a good distance upstream the land on the opposite side of Cedar Fork has been torn and blasted for the dolomite underlying it. One can only wonder what interesting geology now lies crushed on the driveways of southwestern Ohio.

Other Arches in the Peebles Dolomite

Bundle Run Arch (OH-A-ADA-10) — Adams County
Span: 26 feet (7.9 meters); Clearance: 4.6 feet (1.4 meters)

At first glance, Bundle Run Arch (Figure 60), located in the face of a low bluff on Bundle Run in Adams County, appears to be a twin to the Jawbone arches. Both have the appearance of a long, dark opening at the bottom of a low cliff of dolomite above a stream. Both were formed through the joining of an enlarged vertical crevice and a widened horizontal bedding plane. They differ, however, in the configuration of the openings that create their respective arches. Whereas the openings of the two Jawbone arches resulted from incomplete widening of the vertical crevice, the single Bundle Run opening is a widened bedding plane.

The vertical crevice forming the skylight of Bundle Run Arch parallels the face of the bluff for 58 feet and varies from 2 feet wide at the ends to 15 feet wide at the center, although it quickly narrows to half that dimension with depth. The length of the opened crevice is matched by the length of the alcove formed by the widened bedding plane, but only at the center does the alcove extend far enough into the cliff face to meet the vertical crevice. The opening thus formed is

Figure 60. Bundle Run Arch. The view out through the entrance from the skylight. The fallen slab of dolomite is on the left side of the picture.

26 feet wide, less than half the potential span of this arch if the alcove had broken through to the vertical crevice along its entire length. The entrance of this opening facing the stream is 9 feet high at the center, but a rising floor and lowering ceiling shortens its height at the back of the opening to 4.6 feet.

The lintel of Bundle Run Arch varies in width from 5 to 16 feet. The face of the lintel closest to the stream is formed of a projecting cap of solid dolomite which is only 3 feet thick. Farther in, weaker layers survive on the underside of the lintel which rapidly thickens until at the back of the arch it reaches 6 feet in thickness despite the disappearance of the protecting cap of rock.

The back wall of the skylight slopes steeply down to the floor of the arch which itself slopes toward the river at a much gentler angle. No wash appears to enter the arch from this direction and such seeps as were seen were insignificant. With one exception, this arch appears to be the cumulative result of several slow erosive processes that occurred over a great length of time. That one exception, however, is notable. It takes the form of a large, flat rock, part of the same resistant layer that forms the top of the lintel, which apparently once formed the roof of much of the skylight, but which at some point collapsed into it. This tabular slab of dolomite, nearly as long as the front entrance of the arch, separated along the joint plane defining the back wall of the skylight and pitched forward, wedging itself against the rear of the lintel. Much of the skylight behind the arch opening is plugged by this rock and a few smaller ones that were part of the same collapse. While the complete history of this collapse feature is somewhat obscure, it is evident that the vertical crevice was enlarged beneath it since there had to be a void for the rock to fall into. Just such cavities are still found at both ends of the crevice, although the alcoves formed there are only a few feet deep. Since the main block rotated forward and down, the joint plane forming the rear of the lintel must have opened first, removing one line of the block's support and providing space for it to rotate through. Gravity did the rest. Other Ohio arches show evidence of similar collapse, but none of them exhibit the process in so dramatic a fashion.

Old Womans Kitchen, West Arch (OH-A-ADA-14)
— Adams County
Span: 4.7 feet (1.4 meters); Clearance: 2 feet (0.9 meter)

Old Womans Kitchen, East Arch (OH-A-ADA-15)
— Adams County
Span: 6.75 feet (2.1 meters); Clearance: 4 feet (1.2 meters)

Old Womans Kitchen (figures 61 and 62) is an unusual arch — so unusual, in fact, that at first sight it did not appear to meet the definition of an arch at all. The first view is of a dark, wide opening at the base of a rocky bluff rising beyond marshy ground at the junction of two small streams (Figure 62). Across the entrance lies a line of rocks, evidently the former lip of a once more-extended alcove roof which gave way and fell in some distant past time to form this shattered barrier. It is only after entering the spacious alcove beyond that one sees the two openings high on the left side. Framing the sunlit slope beyond, they look like two gigantic eyes gleaming in the darkness. From outside, the two arch openings appear as dark holes at the bottom of a narrow skylight and could, in fact, be considered to form a single opening divided by a bedrock pillar 7 feet wide.

The alcove into which these spectral eyes peer is spacious, having a reach of 36 feet, a height of 9.5 feet and a depth of 16 feet. It is a damp place and one wonders how comfortable a kitchen it would make. Even so, local tradition strongly insists that at one time an old woman lived, or at least cooked, here, running her stovepipe chimney out through one of the holes in the alcove wall. One local informant remembers seeing remnants of the wooden kitchen wall that closed off the front of the alcove when he was a child, although that wall could also have been built by the men who gathered here during the Depression for friendly games of poker — another local tradition. This feature is often called Old Maids Kitchen, but Ohio already has two of those so the name was changed to the one used here.

The alcove and its arches have been eroded into the Peebles Dolomite which shows its usual characteristics of pitted, vuggy surfaces and undercut cliffs in the low bluffs lining the streams of the area.

Figure 61. Old Womans Kitchen. West Arch is on the left; East Arch is on the right.

Figure 62. Old Womans Kitchen. A view showing the large alcove into which the arches open.

Tiffin Arch (A-ADA-11) — Adams County
Span: 12 feet (3.7 meters); Clearance: 4 feet (1.2 meters)

Tiffin Arch (Figure 63), named by the Ohio Cave Survey for the township in which it is located, has a history that differs from that of most of the other arches formed in the Peebles Dolomite. It is one of the few Ohio arches that were formed, at least in part, by cave collapse.

The opening to Tiffin Arch forms a negative feature in a flat upland field covering the top of the resistant Peebles Dolomite. Just to the east, the land plunges steeply into the deep valley of Ohio Brush Creek. Several shallow sinkholes dot the field and are marked from a distance by groves of thin trees. The 400-foot difference between this upland plateau and the river below has provided the fall needed by groundwater to create the sinks. According to the Ohio Cave Survey, none of them leads to a cave of any size.

The sinkhole containing Tiffin Arch is only a few yards from the edge of the cliff. Like the other sinks in the field, it is small, being

Figure 63. Tiffin Arch. A view of the arch from within the collapsed portion of the cave.

approximately 50 feet long and 15 feet wide. Its floor slopes from ground level at the north end to a depth of about 10 feet at the south end. Tiffin Arch spans the center of this narrow depression and appears to be a remnant of bedrock left in place when the roof of a low underground cavity collapsed to form the sink. The lintel is 12 feet long, 6 feet wide and varies between 1 and 2.5 feet thick. Due to the sloping floor of the sinkhole, the clearance rises from 4 feet on the north side of the arch to 5 feet on the south.

The deep valley of Ohio Brush Creek is considered to be a breeched preglacial drainage divide which was cut through by ponded glacial meltwater seeking an outlet during the Illinoian glaciation. Buzzardroost Rock, a striking dolomite promontory on the opposite side of Ohio Brush Creek, represents the site of this divide. It is now part of the Edge of Appalachia Preserve owned and operated by the Cincinnati Museum Center and The Nature Conservancy. A trail leads to its top which offers a fine view of the valley. Since the existence of the sinkhole containing Tiffin Arch is most likely due to the steep gradient between the flat upland where it is found and the river below, it and the arch it contains could not have started forming before the cutting of the gorge of Ohio Brush Creek was well advanced.

According to the owner of the arch at the time it was originally visited, a slab of rock lying on the floor of the sinkhole at the north side of the arch broke away during the 1970s, indicating that collapse as well as solution has played a part in formation of the sink. In spite of its small size, Tiffin Arch is highly instructive, unique in Ohio and surprisingly picturesque in spite of the unfortunate presence of an old farm dump covering the floor of the northern part of the sinkhole.

Castlegate Arch (OH-ADA-12) — Adams County
Span: 11 feet (3.3 meters); Clearance: 9.6 feet (2.9 meters)
Castlegate Arch (Figure 64; Plate 2) has the look of an ancient ruin, as though it were a remnant of the wall of some destroyed city. Such a romantic appearance deserves a romantic name. Castlegate is also a good example of how arches can be overlooked. Although it stands above a heavily traveled highway and is in plain view when the trees surrounding it are leafless, it had not been reported until after this survey began. Then two different land managers in the area

Figure 64. Castlegate Arch. In this view looking through the opening of the arch to the Ohio River, the partially detached block of dolomite resembling the tower of a city wall is outside the photograph to the right.

happened to see it and, suddenly realizing that it was a feature of importance, contacted the Ohio Natural Arch Survey.

The arch is formed in a rough, broken edge of the Peebles Dolomite where the outcrop is cut by the valley of the Ohio River. Here the dolomite forms a bluff about 2 miles long which is broken by steep gullies and wider valleys. The rock itself is vuggy and pitted, and is penetrated by a number of short caves. The whole face of the bluff has the appearance of rotten rock.

Castlegate Arch stands out from the face of the bluff high above the valley floor. The opening pierces a thick fin which is separated from the main body of the bluff by a steep-sided gully which probably began as an enlarged joint. The fin varies in width from 6 to 8 feet. At its very end, beyond the arch, a notch cut down into it from the top separates an 8-foot-high tower-like block, adding further to the illusion of ruined castle walls.

The opening of the arch, which is somewhat rectangular, is 11 feet long and 9.6 feet high. A narrow cave opens into the wall of the gully behind it. Although it does not penetrate far into the hill, the cave emphasizes the fact that the bedrock here has been exposed to weathering and atmospheric erosion for a longer time than in other parts of Ohio. The fin holding the arch is actually a remnant of bedrock which has not yet yielded completely to the ravages of time.

The age of the arch is linked to the age of the valley of the Ohio River at this point, for the arch could not have started forming until the Peebles Dolomite was exposed. Before the advance of the Pleistocene glaciers, this part of North America was drained by a now-vanished major river called the Teays. This master river of the region began in the Carolinas and flowed north through West Virginia into Ohio where it apparently turned westward to flow through Indiana, following much of the route of today's Wabash River, to join the ancestral Mississippi River. During Pleistocene glaciations, the ice sheets blocked this northerly flow and forced the impounded waters to drain more directly to the west, a flow which, after a number of adjustments brought on by several readvances of the ice, resulted in the Ohio River. The area where Castlegate Arch is now located was at that time on or near a divide between the old Portsmouth River which flowed into the main stem of the Teays and a branch of the Manchester-Old Kentucky River system which entered the Teays farther downstream in what is now west-central Ohio (Bray, 1985). Advance of the Kansan or Illinoian glacier blocked the northward flow of the Teays and ponded it. The rising water eventually escaped by cutting across several drainage divides, including the one where the arch is located. This new course, stitched together from bits and pieces of old Teays tributaries, became the present Ohio River.

The bluff in which Castlegate Arch is found would not have existed until the Kansan or Illinoian glaciation, whichever one was responsible for blocking the Teays River. While much work has been done on the history of the glacial derangement of Ohio's drainage systems, much is still in question. However, it is certain that the cutting of this bluff took place before the latest glaciation (the Wisconsinan) and so weathering has had a fairly long time to work on it.

Vertical Dome Arch (OH-A-ROS-03) — Ross County
Span: 1 foot (0.3 meter); Clearance: 4 feet (1.2 meters)

On the upper reaches of a branch of the small stream on which Trimmer and Skull Cave arches are found is a series of low waterfalls. The highest, barely 10 feet tall, falls into a shallow amphitheater weathered into an outcropping of Peebles Dolomite. Although small, the waterfall is notable for the interesting calcium carbonate dome it has built out from the wall behind it over the years.

To the left of the waterfall, a dark, oblong opening 8 feet wide and 4.5 to 8 feet high leads into a chamber 14 feet long and not much wider than the entrance — Vertical Dome Arch (Figure 65). Once inside, however, one becomes aware of a much larger space than expected, for the cavity extends upward at least 15 feet, having the appearance of a tall, thin, ragged dome such as those seen on a larger scale in Kentucky caves. It appears that the interior of the rock bluff into which the cavity has been weathered has been hollowed out — a strangely frequent occurrence in the Peebles Dolomite of the region. Looking back toward the waterfall, the right-hand side of the entrance is seen to narrow into what appears to be an enlarged crevice extending up the height of the dome. In fact, the entire cavity seems to be an enlarged vertical crevice which has been opened into this interesting domed room.

The top of the dome is not shrouded in darkness as might be expected, for a small, unseen opening admits light. A steep climb up the outside of the rock bluff reveals a hole 1 foot wide and 4 feet high. This second entrance qualifies the opening as a natural arch. Since this upper entrance is the most constricted part of the total opening, its dimensions are given as the span and clearance. Looking through this entrance, a third, smaller entrance is seen on the far side of the cavity. This one unfortunately opens out onto a sheer rock face and is not accessible for measurement. It does, however, frame a view of an upper waterfall not quite so high as the one at the lower entrance to the cave.

Figure 65. Vertical Dome Arch. The entrance of this arch, visible as a large, dark hole, opens beside a waterfall. Note the deposits of calcium carbonate on the face of the falls.

Most of the face of this upper falls is formed of the pitted Peebles Dolomite, but the top few feet are thin-bedded Greenfield Dolomite. Although only a few feet thick here, the exposed Greenfield Dolomite is at least 12 feet thick at Skull Cave Arch less than a quarter mile to the north; at Trimmer Arch to the west of Skull Cave Arch it is 16 feet thick. Thus it would appear that the Greenfield Dolomite thins to the south and perhaps to the east, an observation strengthened by the fact that at Baker Fork Gorge in Fort Hill State Memorial several miles farther south, the Devonian Ohio Shale rests directly on top of the Peebles Dolomite and the Greenfield Dolomite has disappeared entirely. Additional observations of outcrops in every direction would be needed to confirm this initial conclusion, but this small series of exposures illustrates how rock units are traced on the ground and eventually outlined on maps. On such seemingly mundane details is the study of geology built.

Greenfield Dolomite

The Greenfield Dolomite, named for the city of Greenfield in northern Highland County, is a drab-colored, fine-grained rock often broken into thin beds 2 to 6 inches thick. In southwestern Ohio, it reaches a thickness of about 100 feet, although this is extremely variable due to its exposure and subsequent erosion after being deposited (Rogers, 1936).

Arches of the Upper Paint Creek Drainage

The upper drainage of Paint Creek is defined as that part above its junction with Rocky Fork. Before the advance of the Illinoian ice sheet, the stream which occupied the present valley of Paint Creek on the border between Highland and Ross counties flowed north. This flow was interrupted by the overwhelming ice. As the Illinoian glacier retreated, it made a stand northwest of the old valley of preglacial Paint Creek. Meltwater from the ice flowed south, cutting much of the present 70-foot-deep gorge into the Silurian dolomites (Rosengreen, 1974). After final retreat of the ice, local drainage formed Paint Creek, utilizing the gorge to flow south and so reversing the former drainage of the area. Since that time, side streams have eroded their own small gorges into the valley wall, cutting their way down to the base level represented by Paint Creek. The three arches presently reported from this drainage, one of which is Vertical Dome Arch described above, are all located on small, unnamed tributaries.

Trimmer Arch (OH-A-ROS-01)* — Ross County
Span: 14 feet (4.6 meters); Clearance: 8.6 feet (2.6 meters)

Trimmer Arch (figures 66 and 67; Plate 3) looks like what most people expect a natural arch to be. Cut through a narrow point of rock bounded by two small streams, its semi-circular opening gives it the appearance of an ancient Roman portal. Its close association with streams also gives it an affinity with the more familiar arches of Utah

Figure 66. Trimmer Arch. This is possibly a meander type of breeched-alcove arch.

and Kentucky, although it, of course, does not approach them in size. The arch spans 14 feet and has a clearance that varies from 8.6 feet on the north to 10.6 feet on the south. This difference is due to a pronounced slope in the floor. In form, the opening is as close to a perfect semi-circle as can be found in any arch in the state. The lintel is 4.6 feet thick at the center.

Several short tributaries flowing into Paint Creek have cut minor gorges of their own in dropping down to the level of the main stream. One of these streams, flowing in from the east, has several branches which occupy narrow, steep-sided, often rock-bound valleys 20 to 50 feet deep. Just before two of the northernmost branches of this small stream join, they run parallel for a short distance, leaving a narrow point of rock between them. The opening of Trimmer Arch pierces this point. The streams here are cutting into both the Greenfield Dolomite and the underlying Peebles Dolomite. The opening of the arch is cut through the Greenfield, but the floor of the opening and the bases of the abutments are in the Peebles. This is very evident when standing below the junction of the two streams and looking at the downstream end of the arch (Figure 67). From this vantage point the massive Peebles Dolomite appears as a solid

Figure 67. Trimmer Arch. A view of the downstream end of the arch showing the underlying massive Peebles Dolomite foundation and the overlying, thinly bedded Greenfield Dolomite in which the opening of the arch is located.

foundation slightly wider than the thinly layered Greenfield Dolomite abutment which sits on it.

The formation of this arch was probably initiated when the streams were at a higher level and opened communication with each other through the point of rock by means of the many bedding planes of the Greenfield Dolomite. The beds have a dip approaching 15 degrees, sloping down to the north, which would indicate that such flow was from the southern side to the northern, although the floor of the arch dips in the opposite direction due to a debris cone which impinges upon it.

The opening of the arch could have formed through chemical and mechanical erosion initiated by water seeping into the bedding planes when the two streams flowed at the level of the arch opening. The process could have included the cutting of alcoves on either side of the point of rock which eventually merged through their back walls. An interesting shallow alcove cut into the side of a neighboring outcrop of Greenfield Dolomite, similar to that in which the arch formed, illustrates such a process. In every way, including size, this alcove resembles the upper part of the arch opening, except that it has penetrated only a few feet into the rock. If Trimmer Arch did form in this manner, then it is Ohio's only known example of a modified meander arch. In this case, the merging alcoves would have been cut by separate streams rather than a single stream making a wide curve.

From the top of the outcrop which contains the alcove, it is apparent that it and the elongated rock mass in which Trimmer Arch is found are parallel and of equal height. The narrow stream which divides them may have been directed at this point by a vertical joint which it widened as it cut down to base level.

The curved top of the arch opening provides our clearest example of an opening which owes its final form to a tension dome. The narrow beds of dolomite forming Trimmer Arch set the stage for just such a process to occur. Here, only a thin slice of the dome remains due to the narrow point of rock in which the opening was cut.

Skull Cave Arch (OH-A-ROS-02)* — Ross County
Span: 7.9 feet (2.4 meters); Clearance: 3.6 feet (1.1 meters)

The enigmatically named Skull Cave Arch (Figure 68) is found a short distance upstream from Trimmer Arch. Surveyed and mapped by the Ohio Cave Survey, it was described in the first issue of *Pholeos*, the journal of the Wittenberg University Speleological Society (Hobbs and Flynn, 1981). Although formed in the same thin-bedded Greenfield Dolomite as Trimmer Arch, it is at the opposite end of the arch-appearance spectrum. Where Trimmer Arch looks just like a natural arch should look, Skull Cave Arch is just that — a cave with three openings. It is these multiple openings that allow it to meet the definition of natural arch. An offset jog in its floor plan 50 feet from the largest entrance blocks incoming light, and were it not for the two additional openings, it would meet the strict definition of a cave as an opening extending beyond the reach of outside light.

135

Figure 68. Skull Cave Arch. This entrance leads to a short cave which opens into the bottom of a sinkhole.

The main entrance opens into a low cliff of thin-bedded Greenfield Dolomite at a point where the stream makes a sharp curve from south to west, following the course that will, within a short distance, bring it to the stream junction where Trimmer Arch stands. This entrance opens into a passage which in most places is 6 to 8 feet high. The smallest constriction of the opening is found at the jog, and its dimensions at this point were taken as the span and clearance of the arch. The opening extends into the hillside a total of 77 feet and then splits into three passages. The smallest one continues in the direction of the main passage, but is quickly pinched off by a dirt fill. A longer passage extends to the right for about 37 feet before it, too, is clogged by a rising floor of dirt. The largest passage turns ninety degrees to the left and runs for approximately 15 feet more, ending at an opening 3 feet wide and 4 feet high at the bottom of a steep-sided, oblong sinkhole 12 feet long, 16 feet wide and approximately 12 feet deep. The thinly bedded Greenfield Dolomite is well-exposed in the sheer rock walls of the sink. A small stream drops from a shallow notch in its southern lip to flow across the floor of the sinkhole, turning to enter the opening of the cave which it utilizes as a grade-level route to the stream at the main entrance.

A third entrance breaks into the upper level of the left-hand wall of the main passage midway between the ninety-degree bend and the three-way split. The jagged opening is 2.8 feet wide and 3.3 feet high, and leads to a steep, debris-covered slope that ends at the cave floor. This entrance opens out into a second sinkhole which is smaller and slightly shallower than the one into which the upstream entrance of the arch opens. It is immediately adjacent to the larger sinkhole and the upper parts of the two sinks have joined together to form a single sinkhole 48 feet long. A steep-sided, dirt-covered ridge extending across the width of the sink and rising to within several feet of its rim marks the dividing wall between the two original sinkholes (Figure 226).

The *Pholeos* article mentioned above states that the cave was rumored to have served as a deposit box for local thieves who stashed their loot in its depths. Whatever ill-gotten gain was hidden there had to be well water-proofed since the small stream flowing through the cave keeps most of the floor flooded to a depth of several inches. Chances are the only things stashed here were rumors and legends. There is no indication in the article as to what sort of skull gave the cave its name.

Cedarville and Springfield Dolomites

Just as much of the surface of Highland and Adams counties is formed on the gently sloping top of the Silurian Peebles Dolomite, so much of the surface of west-central Ohio rests on top of the very similar Silurian Cedarville Dolomite. Like the Peebles, the Cedarville is massive and somewhat susceptible to solution. It and many of the Silurian beds beneath it are, however, more resistant than the underlying Ordovician limestones and shales. Once the resistant Silurian strata were breached by erosion, the exposed Ordovician rocks eroded rapidly. This led to further retreat of the edge of the Silurian outcrop through undercutting and collapse. The eroded edge of the Silurian outcrop in southwestern Ohio is often marked by an abrupt rise in the landscape referred to as the Niagara Escarpment (Figure 9). Over much of its extent, the escarpment is covered by glacial deposits which subdue its topography. In several places, however, glacial meltwater streams pouring over its edge eroded long, narrow gorges into it which are now used by local drainage. Here the underlying rock is exposed in cliffs, resulting in some of Ohio's more dramatic scenery. Clifton Gorge, now utilized by the Little Miami River, is the most famous example. The gorges of Massies Creek and Mad River were formed in the same manner.

While such conditions might seem conducive to the formation of arches and pillars, far fewer are found here than in similar exposures of the Peebles Dolomite. In all cases except Greenville Falls Arch, the features are found where the Cedarville is comparatively thin. They also are formed in part by formations underlying the Cedarville, unlike features in the Peebles which are usually limited to that formation.

This dearth of arches and pillars in the Cedarville Dolomite may be due to its greater resistance to erosion. It may also be a result of erosive forces having less time to work on it than on the Peebles Dolomite. Most of the area in Ohio where the Cedarville is exposed was covered by the Wisconsinan glacier, the last continental ice sheet to enter the state. The cliffs now found in it were formed through rapid down-cutting by glacial meltwater streams which flowed at the end of that stage and then vanished as the edge of the retreating glacier passed north of the continental divide crossing Ohio. This rock has thus been exposed to erosion only since the retreat of the last glacier whereas the exposures of Peebles Dolomite in Highland and Adams counties have been exposed at least since retreat of the Illinoian glacier which preceded the Wisconsinan.

In two of the three arches so far recorded involving Cedarville Dolomite, the opening and abutments are actually formed in the underlying Springfield Dolomite, a much thinner formation with closely spaced bedding planes and very broken rock layers which give it the appearance of a poorly laid brick wall. This makes it very susceptible to removal by erosive processes and results in a distinctive cliff profile where it is found — a sheer cliff of massive Cedarville Dolomite overhanging a recess, which may be shallow or quite deep, cut back into the Springfield Dolomite and floored with more thickly bedded

Euphemia Dolomite which is less susceptible to erosion than the Springfield. Under these conditions, Cedarville Dolomite provides the lintel of the arch and Springfield Dolomite, the abutments. Greenville Falls Arch provides the only exception. In this case, the opening has been made in the Cedarville, although the underlying Springfield plays a very important role.

Tecumseh Arch (OH-A-CLA-01) — Clark County
Span: 2.7 feet (0.8 meter); Clearance: 3.5 feet (1.1 meters)

Tecumseh Arch (Figure 69) is more impressive than the measurements of its span and clearance might indicate. A tabular lintel resting on thin-bedded abutments give it a rugged appearance despite its small size. The arch is located in Mad River Gorge, sometimes called Tecumsehs Shooting Gallery from its association with that famous Shawnee Indian leader. In the late 1700s, Peckuwe, a major Shawnee settlement of the time and one of two reported birthplaces of Tecumseh, was located on the Mad River 2 miles downstream from the arch. One of the largest Revolutionary War battles west of the Allegheny Mountains was fought there when George Rogers Clark and his hastily gathered militia units attacked the town in an attempt to halt Shawnee depredations south of the Ohio River. The site of the town, but not the arch, is now part of George Rogers Clark Park in Clark County.

Tecumseh Arch is composed of a large, tabular piece of Cedarville Dolomite resting on abutments of thin-bedded Springfield Dolomite which, in turn, rise from a foundation of more thickly bedded Euphemia Dolomite. Most of the opening of the arch, which measures 2.7 feet wide and 3.5 feet high, has been broken through the Springfield Dolomite. A short distance upstream, the Cedarville Dolomite has a thickness approaching 40 feet. Nearing the arch, however, its thickness rapidly declines to 7 feet. At the arch, it has an obvious bedding plane which divides it horizontally. The bottom layer has fractured vertically at the site of the arch opening. The top layer also has a vertical fracture which is located farther upslope than that in the bottom layer. These fractures coupled with bedding planes have allowed the top layer of the span and the downslope abutment to move slightly outward, adding a narrow, crevice-like extension to the upper part of the arch opening. The sides of this crevice are mirror images

Figure 69. Tecumseh Arch. The view looking upstream. The cliff face is to the left. The arch opening and the small window in the right abutment have formed in the thin-bedded Springfield Dolomite. The scaled ruler is 12 inches long.

of each other, showing that this portion of the opening is a result of the separation of adjoining blocks in contrast to the unmatched walls of the wider eroded portion of the opening below. Because the vertical fractures in the two parts of the span are offset by several feet, their movement has not resulted in a crevice open to the sky, as has occurred at John Bryans Window (Figure 170).

This movement has been slight, perhaps a foot at most. Exact figures are hard to come by due to the presence of a dump which encroaches on the uphill side of the arch. The movement can be attributed to the weakness of the downslope abutment of the arch which is made of thinly bedded Springfield Dolomite. Its small size (barely 4 feet across) in relation to the heavy span it carries, its position on the downslope side of the arch and the thin layers (3 inches maximum thickness) forming it, which not only weather away rapidly but also allow some horizontal movement along multiple bedding planes, all combine to weaken the abutment. The presence of a second, smaller opening 1.4 feet wide and 1.2 feet high through this abutment certainly does not help its stability.

As a result, the lintel is slowly sliding out toward the valley below. There is a small but noticeable outward tilt to the layers making up the downslope abutment in comparison to those in the upslope abutment. The Euphemia foundation is also tilted downslope, although whether this is due to creep or is a reflection of its original bedding is unknown. In either case, it certainly helps to destabilize the arch. In time, the lintel will slide far enough to overwhelm its support and keep right on sliding to become just another slump block on the lower slope of the gorge.

It may seem that the movement which has already occurred would disqualify Tecumseh Arch from official listing since one of the criteria is that listed arches be a result of the weathering away of bedrock and not of movement. However, in this case the movement has only served to enlarge an opening already made by weathering, and that only slightly.

Tecumseh Arch has formed in the angle of an offset of the upper part of the cliff, which here has the undercut profile typical of cliffs formed in the Cedarville-Springfield-Euphemia dolomites sequence of the region. The back wall of the arch opening is a continuation of the undercut cliff downstream from the arch. The upstream edge of the lintel is formed by a 25-foot offset which meets the cliff at nearly a right angle. This face may be a fracture plane; several enlarged fractures cut into the cliff at the same angle downstream from the arch. In form, Tecumseh Arch is the first in a series that includes Pompeys Pillar (Figure 167) and John Bryans Window which are also formed of Cedarville capstones resting on bases of Springfield and Euphemia Dolomites. In the case of the pillars, the capstone has separated completely from the adjoining bedrock, slightly at John Bryans Window and entirely in Pompeys Pillar. At Tecumseh Arch, separation is imminent, but not complete.

Mad River Gorge, formed as a river carrying glacial meltwater, cut its way back into the Niagara Escarpment during retreat of the Wisconsinan ice sheet. It is the youngest of several gorges which formed in sequence as the retreating glacier exposed ever-lower outlets for its meltwater to follow. While it is possible that the violent floods which created the gorge washed the rock away from the offset in which Tecumseh Arch is found, it is unlikely that the arch itself formed at this time. Its downslope abutment is too delicate to withstand the violence of glacial flooding and the pounding of water-borne rock which accompanied it. And, too, the obvious signs of outward creep visible in the abutment seem to indicate a recent origin for the arch. In geological terms, it appears to be a very young and short-lived feature.

Low Arch (OH-A-GRE-01) — Greene County
 Span: 10 feet (3 meters); Clearance: 2.5 feet (0.8 meter)
Low Arch (Figure 70), found in that part of Clifton Gorge protected by John Bryan State Park, is a prime example of the challenges one can encounter when trying to decipher the story of natural arches in Ohio. Cut through a low projection of the cliff on the north side of the Little Miami River, Low Arch's status as a bona fide arch was initially doubted, mainly because it simply does not look like an arch. However, once a final definition of natural arch was crafted and it was decided to follow it regardless of the aesthetic consequences, Low Arch made the list. Even so, a more recent visit raised other questions.

Low Arch presents the typical profile of the Silurian bedrock found in the region with a heavy lintel of Cedarville Dolomite resting as a cap rock on top of thinly bedded abutments of Springfield Dolomite. It greatly resembles Tecumseh Arch right down to having a smaller opening breaking through the downslope abutment. It is, however, much more massive in form. Although it has roughly the same clearance as Tecumseh Arch, its opening is much wider and deeper, and has a slight curve that gives it more of a cave-like feel. Whereas it is possible for a person, albeit a small one, to step through Tecumseh

Figure 70. Low Arch. The entrance of this arch is marked by the arrow on the sunlit slope on the right. The arrow to the left marks the small window. The scale, to the left of the window, is 12 inches long.

Arch as is expected of any self-respecting arch, one must crawl through Low Arch, and a muddy business it is, too.

Although Low Arch appears to meet the criteria for listing, a few troubling observations must first be resolved. There has been a great deal of rock movement in this area. The top of the arch and the cliff projection in which it is found is littered with blocks of Cedarville Dolomite, most of them ranging between 3 and 5 feet long. A vertical crevice opening into the cliff projection beside the downstream entrance of the arch is filled with smaller blocks. It is possible that this crevice actually splits the lintel, creating a pillar rather than an arch, its true character being masked by the rock infill of the crevice and subsequent soil formation and plant cover over the fill.

An additional problem is posed by the possible downslope movement of this questionable arch. A similar section of rock forming the side of the crevice opposite to the arch opening has definitely moved downslope. The top of its Cedarville capstone and the junction between its cap rock and the underlying Springfield Dolomite are both at a noticeably lower level than are the same markers in the adjoining potential arch. Both markers are also lower than the same rock levels in the nearby intact bedrock cliff. The movement has most likely resulted from slow downhill creep as this block of Cedarville-Springfield-Euphemia dolomites slid on the groundwater-lubricated surface of the underlying shale.

There are, however, several points in favor of Low Arch being reckoned a true arch. One of them is the very fact that the neighboring rock column which has slid downhill is lower than the column forming the arch. If any downhill movement of the arch column has occurred, it must be minor. In fact, the line where the Cedarville Dolomite meets the underlying Springfield Dolomite in the potential arch appears to be at the same level as that in the cliff a short distance away, indicating that no movement at all has taken place.

The cliff also helps to explain the large number of small slump blocks found on top of the arch. The Cedarville Dolomite in this location is divided horizontally into two roughly equal layers. The top layer has several widely spaced bedding planes and is more easily attacked by erosion than is the massive bottom layer. As a result, the top layer has been cut back into a rugged slope marked by ledges of dolomite while the bottom layer remains as a low, sheer cliff. In the area of the arch, this bottom layer extends out from the face of the cliff as a flat-topped projection covering the more fragile Springfield Dolomite underneath. Low Arch is found in the downstream end of this projection. The small slump blocks littering its top are all that remain of the less-resistant upper level of Cedarville Dolomite which has been eroded back off the top of the projection.

As for that rock-filled crevice which might split the lintel of the arch, its trend at the arch entrance is into the cliff projection rather than parallel with the depth of the arch opening. It is filled with soil and rock which prevents following its course with much confidence, but a very shallow trough in the top of the cliff projection may indicate its extent. This trough begins at or near the face of the vertical crevice and curves around the uphill side of the arch, well away from the underlying opening. In addition, the upslope wall of the arch opening itself has the distinctive layered appearance of Springfield Dolomite, not that of a tumbled jumble of broken rock filling a widened crevice. There does appear to have been some very minor downslope movement caused by failure of the thin-bedded Springfield Dolomite beneath the much thicker block of cap rock, but this movement was not responsible for creating the opening of the arch. In fact, the loss of so much underlying Springfield Dolomite to form the arch opening undoubtedly affected its ability to hold the cap rock. Rather than rock movement creating the opening, the creation of the opening has enabled the rock to move.

Such movement as has occurred is extremely limited, however. This is evidenced by a small remnant column of Springfield Dolomite less than a foot thick connecting the floor and ceiling of the opening near the uphill side of the arch. Had the overlying cap rock moved downhill more than a few inches, this pillar would have been deranged in some manner. Instead, its bedding planes remain intact, giving it the appearance of rock left behind by erosive activity.

Just what that activity might have been is another intriguing question. Clifton Gorge, like Mad River Gorge where Tecumseh Arch is found, was cut back into the Niagara Escarpment by a river of glacial meltwater coming off the Wisconsinan glacier. In this case, that river laid the Kennard Outwash, the oldest and highest of the several gravel outwash trains found in the region. It has been traced from the Bellefontaine Outlier in Logan County into the head of Clifton Gorge at the village of Clifton. This glacial river was much larger than the Little Miami River which is now using the gorge. During times of rapid melt-back of the glacier, it is quite possible that the gorge was filled to the rim with roiling, churning, rock-filled water. That would make the gorge a high-energy erosive environment similar to that found on the rock-bound shore of Lake Erie today. Clifton Gorge shows plenty of evidence of this high-energy river. There are remnants of large potholes scoured into bedrock near the top of the cliffs lining it, deep overhangs cut beneath the massive Cedarville Dolomite cap rock and large tumbled slump blocks, several of which stand on end. An especially striking example of this energetic environment is located a few hundred yards upstream from Low Arch where the raging glacial river widened a vertical crevice that separated a massive sliver of dolomite 120 feet long from the cliff behind it. The rushing waters evidently cut away enough of the weak Springfield Dolomite at the bottom of this rock to destabilize it, causing it to lean back, giving the beds forming it a very noticeable tilt approaching forty-five degrees.

It would appear that this churning water was responsible for plucking out blocks of Springfield Dolomite, creating both the arch opening and the "window" cutting through its downslope abutment. This window is positioned so as to funnel water moving downstream into the arch opening, and may have been a major player in creating the opening's unusual width. The broken edges of the Springfield Dolomite do not appear rounded, indicating that the arch opening formed mainly through the plucking away of small blocks of rock rather than through solution or scouring. Because of its location near the rim of the gorge, the arch probably was underwater only during the highest flood stages and so was not constantly exposed to the force of the river. That could explain both the presence of the small interior pillar and the continued existence of the arch itself.

Greenville Falls Arch (OH-A-MIA-01)* — Miami County
Span: 11 feet (3.3 meters); Clearance: 4 feet (1.2 meters)

West of the village of Covington, Greenville Creek has cut a gorge 20 to 30 feet deep into the Cedarville and Springfield dolomites. At the upper end of the gorge the river passes over Greenville Falls, an impressive cascade occupying the full width of the river. Downstream from the falls on the north wall of the gorge is Greenville Falls Arch (figures 71 and 72; Plate 4).

Because the arch is usually approached from the top, its full size is not apparent to most observers. From this vantage point the most noticeable feature is its skylight, a linear opening 7 feet wide and 29 feet long through which the river can be seen almost 30 feet below. Viewed from the river, the arch is more impressive — a massive, overhanging, brooding presence.

This arch is a direct result of erosion by a permanent spring. The Cedarville Dolomite at this point consists of two massive, resistant layers separated by a 5-foot-thick layer of broken, vuggy dolomite. Such layering, while not common, has been noted in the walls

Figure 71. Greenville Falls Arch. A view from above; the spring-cut alcove is to the right. Greenville Creek can be seen through the opening of the arch.

Figure 72. Greenville Falls Arch. A view from across Greenville Creek.

of both Clifton Gorge and Mad River Gorge where it appears as lens-shaped masses up to 100 feet long and 5 feet thick. In several instances these masses have weathered out to form shallow recesses, but, so far as is known, only at Greenville Falls have they resulted in the formation of an arch.

This middle layer of broken dolomite gives free passage to groundwater. At Greenville Falls its flow has been concentrated into a spring with a seasonally heavy output emerging from a hole 1 foot in diameter just behind the arch. The spring once issued from the face of the cliff, but has eroded its way 17 feet back, forming an alcove. A vertical crevice cutting the roof of this alcove parallel to the cliff face has been enlarged to form the skylight, thus isolating the front of the roof of the alcove to form the lintel of the arch. The constant flow of the main spring and other seepages at the same level cause high humidity in the alcove and keep the rocks wet, creating conditions conducive to rapid weathering.

The amount of water coming from the main spring is surprising considering that it issues from the rock just a few feet below the surface of the ground. The drainage is almost certainly local which, given the flatness of the land to the north and the thinness of the level, glacially laid ground moraine which covers it, would not appear to be extensive enough to support much flow at all. Indeed, during some visits, the spring's flow was reduced considerably. However, a ridge-like recessional moraine paralleling Greenville Creek between the villages of Covington and Gettysburg just north of Covington-Gettysburg Road and less than 1,000 feet from the arch provides a covering of till 25 feet thick which no doubt provides most of the water supply. This recessional moraine formed when debris released by the melting ice piled up along the edge of the Wisconsinan ice sheet during a pause interrupting its retreat. A meltwater stream in front of the wasting glacial front may have originally cut the valley and gorge now used by local drainage to form Greenville Creek. The presence of a Late Wisconsinan ground moraine north and south of Greenville Creek and the Late Wisconsinan recessional moraine to the north indicate that the gorge was most likely cut during that same period. The arch could not have begun forming before this time.

The massive lintel actually overhangs the cliff face, making measurements of its span and clearance somewhat difficult. The elliptical opening below it is 10 feet wide and 4 feet high as measured to the level of the skylight floor. The arch itself has no floor unless the river below is considered such. Flow from the spring has caused the lower layer of the Cedarville Dolomite and the Springfield Dolomite below the arch to retreat until they are even with the back of the lintel.

This lintel has a thickness of 10 feet and a width of 8 feet. It appears to be a fine example of an arch surviving because the flow of groundwater through the rock comprising it has been cut off by formation of the skylight. The face of the cliff, weakened by continued groundwater flow through it, has retreated on either side of the lintel, leaving the arch projecting and suspended over thin air. A 5-foot-thick layer of permeable Springfield Dolomite underlying the Cedarville cap rock at river level has aided this retreat. Abundant seepage from this layer has created an undercut at the base of the cliff which encourages collapse of the cliff face. Once this retreat reaches the abutments attaching the arch to the cliff, the arch will fall.

The locally unusual conditions provided by the gorge support some equally unusual plants. Just downstream from the arch, seepage from the upper levels of the rock face supports a small colony of plants more commonly found in the alkaline wetlands called "fens" scattered across west-central Ohio. Shrubby Cinqufoil (*Dasiphora fruiticosa*), Limestone Savory (*Calamintha arkansana*) and Wand-lily (*Zigadenus elegans*) can be found in this interesting "hanging fen" clinging to the cliff face. Harebell (*Campanula rotundifolia*), a delicate blue western wildflower dotting the shaded, moist depths of the gorge is here living at the easternmost edge of its natural range. The upland behind the arch supports a remnant prairie community with Butterfly Weed (*Asclepius tuberosa*), Nodding Wild Onion (*Allium cernuum*) and Gray-headed Coneflower (*Ratibida pinnata*).

Human activity has deeply impacted this area. Water power provided by the falls was harnessed during the early phase of Euro-American settlement to support a growing industrial complex. The remains of a penstock and other ruins directly across the river from the arch mark the site of an electricity generating station which was the last facility to use the river for power. Greenville Creek is part of the Stillwater State Scenic River system and this interesting place is now contained within Greenville Falls State Nature Preserve. Greenville Falls Arch is the westernmost arch in Ohio with the exception of a small one found just downstream from it. Unfortunately, this arch has been filled with glacial boulders removed from the farm field behind it and could not be studied.

Arches in Other Silurian Formations

Lions Den Arch (OH-A-PIK-01) — Pike County
Span: 9.5 feet (2.9 meters); Clearance: 4.75 feet (1.4 meters)

The valley of Sunfish Creek near the village of Byington is something of an anomaly. Although well within the sandstone and shale hills of the Appalachian Plateau, the creek has eroded into the Monroe Limestone below and runs between bluffs of calcareous rock, some of them approaching 30 feet in height. The result is a window of karst terrain containing several caves and Kincaid Springs, one of the largest springs in Ohio.

On the south shore of Cave Lake, the artificial center-piece of a private campground within this karst window, yawns the impressive opening of Frost Cave, 26 feet high and 55 feet wide. Like most caves in this region, the impressive opening rapidly shrinks to an impassible crevice. Across the lake is much smaller Lions Den Cave, named by the Ohio Cave Survey for the many ant-lion dens they found in its entrance. These predacious insect larvae rest at the bottom of conical pits they build in dry sand or dust wherein they wait for hapless victims to slide down toward their waiting jaws.

At the opposite end of the bluff containing the cave is Lions Den Arch (Figure 73), named after the cave. The arch is not large, having a span of 9.5 feet. The thickness of the lintel varies from 3.5 to 1.5 feet thick; it is 3.3 feet wide at its narrowest point and 4.75 feet at the widest. A skylight 8 feet wide and 10 feet long opens behind it. The back wall of the skylight merges with the floor of the arch in a steep slope toward the lake below. This gives the arch a clearance at the front of 4.75 feet.

Figure 73. Lions Den Arch. A view of the arch from below. The scale is 12 inches long.

This arch is similar in form to, although much smaller than, Spring Creek Arch in Fort Hill State Memorial. Both have the appearance of being erosional remnants of rock "rotting away" beneath the onslaught of the elements.

Ohioview Arch (OH-A-ADA-13) — Adams County
Span: 7.3 feet (2.2 meters); Clearance: 1.4 feet (0.4 meter)
Ohioview Arch (Figure 74; Plate 5), located in the rim of a 30-foot-high bluff and overlooking the Ohio River a half mile away, is a bit more complex than most arches in the state. It has multiple openings and its form can best be described as a pillared alcove with a skylight. Given the striking ochre color and smooth surface of the rock, it appears to have formed in the Bisher Formation, a prominent ledge-forming rock of Silurian age in this part of Adams County. The bluff in which it is found is part of a discontinuous line of low cliffs marking the top of the ridge which forms the north side of the Ohio River Valley at this point.

There are two entrances to the arch which open at nearly right angles to each other and which are separated by a blocky pillar 13.2 feet wide. The west entrance is 9.4 feet wide and 5 feet high; the east

Figure 74. Ohioview Arch. A view of the west entrance; the skylight, marked by an arrow, is in the upper left corner of the picture beneath the tree.

entrance is 10.8 feet wide and 4.25 feet high. The span of the arch, measured at the most constricted point of the opening between the two entrances, is 7.3 feet, and the clearance, measured at the same point, is 1.4 feet. The constriction is due to a fan of soil and debris washed through the skylight at the back of the arch. This narrow crevice 11 feet long and 6 feet wide opens parallel to the west entrance and defines the lintel which is 20 feet wide and 4.6 feet thick. The back wall of the skylight is very steep, rising at an angle greater than 45 degrees. Below the lintel, the slope lessens where the rock wall at the back of the skylight is masked by a cone of debris washed in off the hillside above.

The depth of the opening between the west entrance and the skylight forms the main axis of the arch which is met at a right angle by the opening leading to the east entrance. The east wall of this opening is a continuation of the back wall of the skylight. The interior walls of the arch are fairly smooth and clusters of small, grape-like botryoidal growths of precipitated calcium carbonate are scattered across the ceiling. The outer 2.5 feet of the west wall of the eastern entrance is separated from the rest of the bedrock between the entrances by a vertical crevice which has been widened into a lopsided triangular window 1 foot wide at the bottom and 2.5 feet high. The crevice which formed it can be traced across the ceiling of the arch in a line nearly parallel with the plane of the eastern entrance and down the east wall of the opening where it has been widened into a rectangular alcove extending 1.5 feet into the rock.

The dry, dusty floor of the arch is liberally sprinkled with sharp rock fragments which make their presence known when one is crawling around trying to get measurements. They do not appear to have broken from the smooth walls or ceiling and may have washed in through the skylight. The larger blocks of rock scattered about and nearly buried by the debris, however, are more difficult to explain. One flat slump block over 5 feet wide lying in the eastern opening is made of granular crinoidal hash formed entirely of tiny circular plates which are the disjointed remains of sea-lily stems jumbled together along some Silurian beach.

Ohioview Arch is a peaceful place, high on its isolated ridge surrounded by aging woodlands. Most amazing is its undisturbed state — no trash, no graffiti, no campfire stains. Just the rock, the wind, a soaring vulture and piles of raccoon scat. It is part of a small piece of ground which has recently, and fortuitously, become an as yet unnamed state nature preserve. It will remain closed to public visitation until adequate access and protection can be provided.

Needles Eye (OH-A-OTT-01)* — Ottawa County
Span: 3 feet (0.9 meter); Clearance: 9.5 feet (2.7 meters)

It is appropriate that Ohio's most famous example of a coastal arch should be located on our inland sea. Needles Eye (figures 75–77) has been a landmark in the harbor at Put-in-Bay on South Bass Island in Lake Erie for well over a century. This arch pierces a fin of thick-bedded Put-in-Bay Dolomite thrusting out from the northeast end of Gibraltar Island which lies at the entrance to the harbor. The narrow opening of this arch is visible to sharp eyes from the western seawall protecting Perrys Victory and International Peace Memorial in the village of Put-in-Bay.

Figure 75. Needles Eye. This recent photograph shows the arch as it appears from the harbor of Put-in-Bay.

The opening of the arch is quite angular and obviously joint controlled. It has a span of 3 feet and a clearance of 9.5 feet from the lake bed. Of this height, 4.5 feet extended above the level of the lake at the time it was measured. This figure is, of course, subject to change as the lake level fluctuates. The width of the lintel is 3.5 feet although the rock fin it pierces is somewhat wider. All measurements were made against the northeast face of the arch from a small boat held in its opening by the strong arms of Nate Fuller, geologist with the Lake Erie section of Ohio DGS. The measure of clearance was the most difficult to secure and is the least accurate.

A photograph of Needles Eye published early in the 1900s (Figure 76) and a picture made late in the nineteenth century for a souvenir booklet (Figure 77) show that the feature has remained virtually unchanged since that time. Such stability is unusual in the wave-battered environment of the Lake Erie shore and is due to the arch having formed in the tough, massive Put-in-Bay Dolomite. Quite possibly the arch was there when Oliver Hazard Perry kept a lookout on the high point of Gibraltar Island above it to watch for the approaching British fleet during the War of 1812. When the British finally

Figure 76. Needles Eye as it appeared about 1900 (Van Tassel, 1901). The viewpoint is the same as that in Figure 75.

arrived, they met the hastily formed American naval forces in the Battle of Lake Erie. The Americans won, which is why this is a study of the natural arches and pillars of Ohio and not of Ontario.

Toward the end of the nineteenth century, Gibraltar Island along with its Needles Eye was purchased by Jay Cooke, a wealthy financier originally from Sandusky. To Mr. Cooke goes much of the credit for financing the Union side of the Civil War. On Gibraltar he built The Castle, a crenellated vacation home where he enjoyed entertaining visitors. Boat trips around the island to see Lovers Cave and Needles Eye were part of the program.

Today Gibraltar Island and Jay Cooke's castle are owned by The Ohio State University and provide a base for Stone Lab, the school's biological field station on Lake Erie.

LeMarin Arch (OH-A-OTT-02) — Ottawa County
   Span: 4.5 feet (1.4 meters); Clearance: 6 feet (1.8 meters)
   For an arch to last long in the energetic environment of the Lake Erie shore, it would have to form in the island archipelago or on the

Figure 77. Needles Eye as shown in a nineteenth-century tourist souvenir booklet. This view of the lake side of the arch looks toward Put-in-Bay harbor. (Use courtesy of Lake Erie Islands Historical Society)

coast of Marblehead Peninsula where resistant limestones and dolomites occur at the surface. The rest of Ohio's share of the lakeshore is fronted with weak shales or even weaker unconsolidated glacial till. With this in mind, a brief reconnaissance of South Bass and Kellys islands and the western shores of Marblehead Peninsula was undertaken. A number of sea caves were noted, as was an interesting gravity arch formed when a pillar collapsed. However, only three true arches were found.

LeMarin Arch (Figure 78), located on Marblehead Peninsula near the lakeside development of the same name, is a pillared alcove. The rise of rock strata from east to west in this part of Ohio brings the more-fractured and therefore weaker Tymochtee Dolomite to the surface on the west side of South Bass Island and Marblehead Peninsula where it forms the lower part of the lakeshore cliffs. This puts the weaker stone at the level of the most intense wave action. Erosion in this layer can be extremely rapid.

Figure 78. LeMarin Arch. The arch opening breaks through behind the rock buttress in the center of the picture.

The alcove of LeMarin Arch has been eroded into this layer of rock at lake level. The column dividing the opening of the alcove may have been isolated by the widening of a vertical fracture between it and the back wall of the alcove. The floor of the arch was barely above water at the time it was visited. Higher lake levels in the past have no doubt contributed greatly to its formation.

North Sugar Rock Arch (OH-A-OTT-03) — Ottawa County
Span: 2.5 feet (0.8 meter); Clearance: 4 feet (1.2 meters)

South Sugar Rock Arch (OH-A-OTT-04) — Ottawa County
Span: 12 feet (3.7 meters); Clearance: 6 feet (1.8 meters)
The two Sugar Rock arches (figures 79 and 80) are found at opposite ends of a sea cave eroded by the waves of Lake Erie into the base of Sugar Rock, a dolomite headland on the west side of Marblehead Peninsula. The alcove is 53 feet long and extends 25 feet

Figure 79. South Sugar Rock Arch and North Sugar Rock Arch. South Sugar Rock Arch opens at lake level at the far right side of the alcove; North Sugar Rock Arch opens through the buttress, marked by an arrow, at the left end of the alcove.

into the 15-foot-high cliff. It is terminated at each end by rock buttresses which are pierced by the arches. The opening in the north buttress, about 10 feet thick, has a span of 2.5 feet and a clearance of 4 feet. The south buttress is only 8 feet thick, but the opening piercing it has a 12-foot span and 6-foot clearance.

The landward walls of both arches continue the back wall of the alcove and represent a joint or other line of weakness paralleling the cliff face at this point. The Sugar Rock arches and LeMarin Arch will

Figure 80. South Sugar Rock Arch. The arch opening as viewed from within the alcove.

not last as long as Needles Eye. Their foundations in the broken Tymochtee Dolomite are too weak to withstand the pounding of Lake Erie's waves for any length of time.

Arches in Devonian Strata

Ohio's Devonian rocks are found in a narrow band reaching from the Ohio River near Portsmouth to Lake Erie at Sandusky, and then east along the lake shore into Pennsylvania (Figure 7). They are also found in a curving band across the northwestern corner of the state and, as an interesting outlier resting on Silurian rock, in Logan and Champaign counties. The oldest Devonian rock exposed is the Columbus Limestone which contains fossils of clams, brachiopods, corals and some of the earliest known fish. Arches have been found in the Columbus and overlying Delaware limestones. Later in Devonian times, conditions changed dramatically and the dark Ohio Shale was deposited.

Dublin Arch (OH-A-FRA-01) — Franklin County
Span: 8 feet (2.4 meters); Clearance: 6.75 feet (2 meters)

In northern Franklin County a small stream enters the Scioto River from the east. In falling to the level of the river, it has cut a

159

gorge which, although small, possesses great geologic and scenic interest, for it has cut down into the Columbus Limestone, the same rock which has given Ohio most of her commercial caves.

In keeping with this characteristic, the sinuous gorge of this tiny stream contains several karst features. At its head, the water drops consecutively into two circular, high-walled basins which fill with spray and roar during flood episodes. Below the waterfalls, the limestone floor of the stream is seamed with joints widened by solution which lead to secret channels into which some of the water carried by the stream finds an underground route to the Scioto. One of the largest of these "swallow holes" is 2 feet long and 6 inches wide.

Several dark openings are also found in the gorge walls, two of which can be entered for short distances. One of these small caves has been breached by the stream where it curves against the north wall of the gorge. Although the original passage was only 3 feet high as determined from the section of cave which remains, the breached part has been enlarged by weathering into an arch-like expanse 10 feet high set against the wall of the gorge. Seven feet of limestone remains between the top of this blind arch and the top of the gorge. A section of this overhang has fallen through behind the rim of the gorge wall to form the skylight of Dublin Arch (Figure 81).

Dublin Arch is a medium-sized arch for Ohio, having a span of 8 feet and a clearance of 6.75 feet. The lintel is 7 feet thick and varies in width from 1.5 feet to 3.5 feet. The back wall of the skylight slopes steeply down to a narrow ledge forming the floor of the much larger blind arch 13 feet below.

If any arch in Ohio is an example of joint controlled form, it is Dublin Arch. Joint control is very obvious in the angular outlines of both the lintel and the 7-foot-wide skylight. To date, no other surface arches have been reported from the Columbus Limestone in spite of its marked tendency to form karst terrain. This is no doubt due to the fact that most of the outcrops of Columbus Limestone found in Ohio are in the glaciated part of the state and are now hidden beneath a thick covering of glacial till.

Figure 81. Dublin Arch. The small skylight of this arch is visible in the foreground.

Olentangy Caverns Arch (OH-A-DEL-01)* — Delaware County
Span: 2.25 feet (0.7 meter) Clearance: 3 feet (0.9 meter)

Arches are surprisingly common in caves. They form in the same ways that surface arches form, the only difference being that they are underground. Olentangy Indian Caverns, a commercial cave near Delaware, offers two readily accessible examples: Olentangy Caverns Arch (Figure 82) and Leatherlips Arch (figures 83 and 84). Olentangy Caverns Arch is located in the Indian Council Chamber, named for a table-like rock in the center of this enlarged passageway which is thought to have been an Indian meeting place. When discovered in 1821, the cave was partially filled with sediment. This was cleared out between 1934 and 1936 to enable the cave to be opened to visitors. During the excavation, a number of flint projectile points were found around Council Rock, leading the owners to assume it had been used by Indians.

The thickness of the sediment that was removed is not known, but presumably the surface of Council Rock was above it. If so, then

Figure 82. Olentangy Caverns Arch. This view is of an underground arch. Note the offset section of the narrowest part of the frame on the left side of the opening.

Olentangy Caverns Arch was also above the surface of the fill, for the lowest part of its opening is at nearly the same level as the surface of Council Rock. With a clearance of 3 feet, it barely meets the size requirement for listing. It apparently formed through solution of the Delaware Limestone bedrock by the same acidic groundwater responsible for enlarging the crevices forming the cave passages, and is therefore considered to be a bedrock-texture arch. A pair of horizontal bedding planes in the narrowest part of the frame encircling the opening has allowed the short section of rock between them to slip sideways, producing a 1-inch offset. It is unlikely that gravity alone was the cause of this shift since the bedding planes have no visible slope. These planes can be followed around the walls of the Council Chamber and have isolated similar blocks of bedrock at several other places. Some of these exhibit the same offset noted in the frame of the arch. One possible explanation for the shift is an earthquake. Although fairly rare, earthquakes do occur in Ohio. A particularly severe jolt, especially if it resulted in a sideways motion, could have shifted these blocks.

It is presumed that Olentangy Caverns Arch formed at the same time as did the cave passage in which it is found. Similar solution cavities and sinkholes buried beneath glacial till have been discovered by quarrying operations in the region. These caves are filled with residual clay and chert resulting from solution of the surrounding limestone. Since no glacially transported igneous or metamorphic rocks are found in this fill, the caves are thought to have developed before the onset of glaciation (Stauffer, *et al.*, 1911). The composition of the fill removed from Olentangy Indian Caverns is not known, but can reasonably be assumed to have been similar to that found in the other solution features in the area. If so, then this cave could have formed during the Tertiary Period, making the arches it contains among the oldest found in the state.

The caverns were originally discovered by J. M. Adams who was looking for a missing ox. He found the unfortunate animal dead at the bottom of the sinkhole leading into the cave. Later, a 16-year-old stagecoach robber named L. M. Wells carved his name and the date 1834 on the wall of the Indian Council Chamber. Wells, as unfortunate as the ox, eventually was caught and hanged for his crimes.

Leatherlips Arch (OH-A-DEL-02)* — Delaware County
Span: 1.6 feet (0.5 meter) Clearance: 5.25 feet (1.6 meters)

Although several arches were noted in Olentangy Indian Caverns, only two were found to be large enough for listing. Leatherlips Arch (figures 83 and 84) is a tall, narrow opening found near a rocky profile said to resemble the face of Chief Leatherlips, a leader of the Wyandot Indians who was executed nearby on the orders of Tecumseh's brother The Prophet in 1810. Chief Leatherlips was too much a friend of the White Man to suit these two Indian leaders who were striving to drive all Euro-American settlers out of the Ohio country.

Leatherlips Arch is a bedrock-texture arch, formed through the solution of the Delaware Limestone by acidic groundwater. This is the same process that enlarged the vertical crevices forming the cave passages.

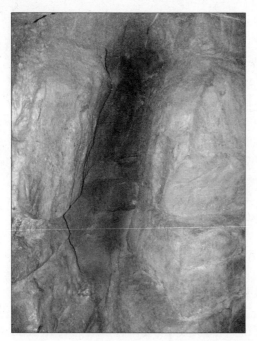

Figure 83. Leatherlips Arch. This is another underground arch found in Olentangy Indian Caverns.

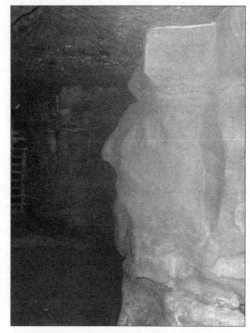

Figure 84. The rocky profile of Chief Leatherlips in Olentangy Indian Caverns.

Arches in Mississippian Strata

Mississippian rocks in Ohio include shale, siltstone, sandstone and conglomerate, all formed from debris washed into an inland sea from high ground to the north and mountains to the east. They outcrop from Scioto County on the Ohio River north to Lorain County and then east into Pennsylvania, paralleling the exposure of Devonian-aged rock. They are also found in the northwestern corner of the state (Figure 7). Although arches and pillars are found in several Mississippian rock formations, the greatest number occurs in the Black Hand Sandstone.

Black Hand Sandstone

Some of Ohio's most spectacular scenery is carved from the Black Hand Sandstone member of the Cuyahoga Formation. This important bed had its origin some 350 million years ago in highlands that stood southeast of what is now Ohio. As erosion wore these highlands down, the debris was carried northward by a major river and dumped into the sea which covered Ohio at that time. This mass of river-tumbled sand and pebbles formed a delta much like the one developing today in the Gulf of Mexico at the mouth of the Mississippi River. The cross-bedding and pebble lenses of the Black Hand Sandstone reflect this origin (Hansen, 1975).

Iron compounds eventually cemented the debris together, solidifying it into rock. These compounds give the Black Hand Sandstone its warm, brown color and impart a greater resistance to erosion than is found in most neighboring rock strata. The result is that areas underlain by Black Hand Sandstone tend to stand higher than surrounding areas where the bedrock is less able to withstand the onslaught of weathering. This is especially noticeable where the Black Hand is found in the form of a massive conglomerate. One such place is the gorge of the Licking River near Newark where a cliff of Black Hand Sandstone once bore an Indian pictograph depicting a black hand — the source of this sandstone's peculiar name.

The greatest development of the conglomerate phase of the Black Hand Sandstone is found in western Hocking County. Here it reaches a thickness approaching 250 feet. Streams cutting into it occupy narrow, steep-walled gorges and head in box canyons. Cliffs and waterfalls are common. Here in the area commonly referred to as the "Hocking Hills" are found some of Ohio's most popular state parks, forests and nature preserves. Here also are found most of the arches and pillars reported from the Black Hand Sandstone.

The arches and most of the other scenic features of the region owe their existence to the layered structure of the Black Hand Sandstone seen here. Differential cementation of the grains making up the rock has resulted in a soft, easily eroded layer sandwiched between two harder, cliff-forming layers above and below. Rapid weathering of the less-resistant middle layer from between the tough upper and lower layers results in the large alcoves, locally called "caves," which are so numerous in the region. Ash Cave is one of the most famous examples (Figure 85). Where streams meet the edge of the resistant upper or lower layer, they drop over, forming the many ephemeral waterfalls that add so much to the scenic beauty of the region.

Of major importance to the formation of the natural arches and pillars found here is a series of near vertical joints cutting the Black Hand Sandstone. There are two major sets which intersect each other at roughly right angles, one set trending north-south and the other east-west. Their effect can be seen in the angular walls of the gorges and is often displayed as flat planes defining one or more walls of an arch or sides of a lintel.

As important as joints and the layered construction of the Black Hand Sandstone has been in forming the natural arches of the Hocking Hills, these two characteristics cannot explain all of the arches found here. The area contains a surprisingly large number of pillared-alcove-type arches which appear to owe their existence to minor zones of weakness and strength within the three major layers making up this formation. Differences in cementation within each layer add to the complexity of the landscape as more-resistant knots and ledges of stone are brought into relief by the removal of less tightly cemented rock from around them. This removal often forms alcoves. If a vertical band of more-resistant rock happens to form part of the cliff face where alcoves are being hollowed out, it will remain as a vertical lintel, creating a pillared-alcove arch. Most of the arches attributable

Figure 85. Ash Cave in Hocking Hills State Park. This spectacular alcove, one of the largest in the state, has been carved into the soft middle layer of Black Hand Sandstone.

to this process are small to medium-sized, but some, such as Early Arch, can attain respectable dimensions.

Hocking County takes its name from the Hocking River which flows through it. The original name of the river was Hockhocking, a Delaware Indian word meaning "bottle." A few miles above Lancaster, the river runs in a straight, narrow channel for a short distance and then plunges over a waterfall below which the gorge widens, giving it the appearance of a bottle when viewed from above. The gorge is cut into the Black Hand Sandstone and contains several interesting features. Although no arches have been reported in this gorge, it is crossed by a covered bridge and sports an impressive old wooden mill on its edge. Rock Mill, claimed to be the oldest and largest grist mill in Ohio, was once in danger of collapse. Fortunately, it was reprieved at the last minute by Fairfield County Parks and is being restored, insuring that this picturesque example of the human use of Black Hand Sandstone geology will continue to grace the landscape.

Arches of the Hocking River Drainage

Rockbridge (OH-A-HOC-01)* — Hocking County
Span: 92 feet (28 meters); Clearance 40 feet (12.2 meters)

Although only two arches have been reported from the Hocking River drainage, one of them is major by any standard. With a span of 92 feet, Rockbridge (figures 86 and 87; Plate 6) is the longest natural bridge in Ohio. It is also the only natural arch in Ohio with a town named after it. Located at the head of a picturesque gorge leading up from the Hocking River, it has a long history as a tourist attraction (Figure 1). Completion of the Hocking Valley canal paralleling the river made the natural bridge easily accessible, and the construction of the Columbus, Hocking Valley and Toledo Railway along the same route later in the nineteenth century continued its popularity as a picnic site into the early 1900s. Photographs taken of it at this time show that the sheltering forest surrounding the natural bridge had been heavily timbered, making it readily visible from the railroad (Figure 87).

With abandonment of the canal and railroad, the bridge lost its accessibility and sank into semi-oblivion. As the forest thickened around it once again, many residents of the nearby town of Rockbridge even forgot that their village was named after a natural feature, believing that the name came from a bridge constructed across the Hocking River. In 1978, 49 acres surrounding the bridge were acquired by the state and dedicated as Rockbridge State Nature Preserve, assuring its continued protection and making it once again a destination point for the curious.

Rockbridge originated as a typical alcove carved into the soft middle layer of the Black Hand Sandstone at the head of a short, steep-sided valley cut by a small, unnamed tributary of the Hocking River. Three intersecting vertical fractures in the roof of the shelter

Figure 86. Rockbridge. This is Ohio's longest natural bridge.

Figure 87. By 1900, the old-growth forest around Rockbridge, like most of the forest throughout the rest of Ohio, was long gone (Van Tassel, 1901).

were gradually enlarged through weathering until the block they surrounded was no longer adequately supported and fell, creating a skylight. Much of this weathering was no doubt due to water from the intermittent stream seeping down into the fractures when its course led across the roof to the rim of the shelter. The angular configuration of the skylight clearly indicates joint control. No large sandstone blocks from the collapsed roof are found on the floor of the shelter, indicating that the rock most likely shattered on impact. The stream which now falls through the skylight would have caused rapid disintegration of any blocks which survived.

The bridge itself is formed of the remaining front rim of the alcove roof. It is 3 feet thick and 6 feet wide near its center. Both dimensions increase toward the ends of the lintel, the thickness slightly and the width dramatically, reaching 26 feet at the west end. There is a 40-foot drop from the top of the lintel to the plunge pool which has been cut 10 feet deep into the hard lower layer of the Black Hand Sandstone forming the floor of the shelter. The part of the alcove which retains its roof is itself impressive, being almost 40 feet deep. Before the ceiling collapsed, the shelter was nearly 65 feet deep.

The small gorge leading down to the Hocking River from Rockbridge is a geologically recent addition to the landscape. Before the Illinoian glaciation, the ancestral Hocking River, part of the Teays River drainage system, flowed to the northwest through the broad valley now utilized by US Route 33, south of the present course of the river. During the advance and retreat of the Illinoian glacier, the ancient valley of the river became an outlet for vast amounts of southward-flowing meltwater which carried sand and gravel released from the melting ice front. So much glacial debris was carried into the old valley that it was filled to the rim with only a few bedrock knobs protruding above the surface. This allowed the glacial outflow to meander across divides between the original valley and those of several former tributaries.

When the Illinoian glacier finally retreated for good and the present Hocking River began incising its southerly course into the thick glacial fill of the now-buried ancient valley, the river found itself entrenched in the tributary valleys it had stitched together and was forced to cut across the old bedrock divides between them. That is the origin of the narrow, steep-walled part of the present valley between Rockbridge and Logan (Baker, 1965). The small stream responsible for Rockbridge is located on one of the bedrock knobs rising above the valley fill of Illinoian outwash deposits. Its origin may date back to Tertiary time and the old Teays River, but the gorge it occupies is much younger, having been cut to reach the new base level represented by the modern Hocking River. The natural bridge could not have started forming until this gorge and the alcove at its head were in existence which would not be before late Illinoian time.

Hintz Hollow Arch (OH-A-FAI-01) — Fairfield County
Span: 3.3 feet (1 meter) Clearance: 1.5 feet (0.5 meter)

The small Hintz Hollow Arch (Figure 88) is found in the valley of an unnamed tributary entering the Hocking River from the west a short distance south of Lancaster. Atmospheric weathering has broken through the end of a narrow slump block 28 feet long to form the opening. The slump block has tilted away slightly from its parent cliff and may even have slid downhill a short distance, but the movement has had little effect on the arch opening other than perhaps making it more exposed to atmospheric weathering. This portion of the rock is pierced by several holes, but only Hintz Hollow Arch is large enough to qualify for listing.

The notable susceptibility of the rock to weathering at this point may be due to weaker cementation of the sand grains and the thinner width of the rock at this end of the slump block. It is doubtful that the arch started forming before the vertical crevice separating the slump block from its parent cliff had widened sufficiently to allow atmospheric weathering to attack it from both sides. Nearby, the cliff presents an outstanding exhibit of such differential erosion where weaker sandstone has been eroded back 8 to 12 inches exposing more heavily cemented layers that jut out from the face of the cliff like thin filigreed shelves.

Of special interest here is the large and healthy population of Great Laurel or Rosebay (*Rhododendron maximum*) found all along the cliff line in masses that are in places so thick that it is difficult to find a way through. This large shrub, reminiscent of the southern

Figure 88. Hintz Hollow Arch. This small opening has formed in the end of a large block of Black Hand Sandstone which has broken away from the cliff located behind the viewer.

Appalachian Mountains, is thought to have made its way into Ohio along the sheltered valley of the preglacial Teays River which headed in what is now western North Carolina. The northern section of the river was covered by the advancing continental ice sheet, isolating sheltered colonies of Great Laurel and other southern species from the main populations to the south. Great Laurel reaches the northern limit of its range in this part of the Hocking River Valley and is listed as a threatened species in Ohio.

Arches of the Queer Creek and Pine Creek Drainages

These two streams are responsible for some of the wildest scenery found in the state of Ohio, scenery that attracts thousands of

172

visitors every year and supports a multi-million-dollar recreation industry. When most people talk of vacationing in the Hocking Hills, this is the region to which they are referring. Its allure has long been recognized and some of the earliest conservation lands purchased by the state were located here. Now it is home to the multi-unit Hocking Hills State Park which contains Old Mans Cave, Ash Cave, Cedar Falls and Rock House among other areas. Here, too, are found Conkles Hollow and four other state nature preserves as well as Hocking State Forest and a multitude of private campgrounds, cottages and other recreation amenities.

The foundation of all this enterprise is the Black Hand Sandstone which here reaches its greatest thickness, but the rock itself would be a poor attraction without the action of Queer Creek and Pine Creek, two seemingly inconsequential streams that have cut their way deeply into it, creating the impressive gorges, hollows, alcoves and "caves" that give the area its wild, scenic character.

Here Ohio's largest regional grouping of natural arches is found. The vast number of amphitheaters and box canyons containing long reaches of sheer rock cliffs scattered over several square miles provides all the conditions required for arch formation on a scale unequaled anywhere else in the state. More natural arches have been reported from these two small drainages than from any other region in Ohio, and were the area to be systematically and thoroughly explored, that number would undoubtedly be greatly expanded.

The history of the formation of the Hocking Hills arches is intimately tied to the history of the gorges, or "hollows" as they are called locally, in which they are found. The courses followed by Queer and Pine creeks and their tributaries draining these many hollows may be some of the oldest continuously occupied valleys in the state, dating back to Teays River times. The obvious dendritic pattern of these drainages, like branches extending out from a tree trunk, shows that little or no glacial derangement has taken place here. That is not to say, however, that these streams were totally unaffected by the massive sheets of ice that advanced and retreated several times over the Midwest.

During the Teays stage, Pine and Queer creeks flowed into Laurelville Creek, a northwest-flowing preglacial stream that cut across present Pickaway County to meet the larger Groveport River south of Columbus (Stout, Ver Steeg and Lamb, 1943). This river in turn emptied into the main stem of the Teays.

During one of the Pleistocene glaciations, many of the streams in Ohio cut their valleys to a lower level than they presently occupy. During this Deep Stage Drainage, Pine and Queer creeks flowed into Adelphi Creek which ran west to join the Newark River above Chillicothe. The Newark River in its lower reaches flowed south, following much the same course as the present Scioto River, and joined the old Pomeroy River at Portsmouth to form the Cincinnati River which, with a few exceptions, followed the course of what is now the Ohio River. It was probably during this Deep Stage that the valleys of the upper reaches of Pine and Queer creeks and their tributaries were deepened as the water cut its way down to the lower base level represented by the Deep Stage streams and formed the hollows seen today.

Later glacial advances filled most of the Deep Stage valleys with glacial debris, raising their floors to their present levels. This infilling did not reach into the higher parts of Pine and Queer creeks which still flow over solid rock beds. It did, however, affect Adelphi Creek, which was blocked and ponded. The Illinoian ice, which advanced farther into this region than the ice of any other glacier, reached the present junction of Pine and Queer creeks at the village of Haynes. The resulting blockade created a lake that flooded 4 miles of the Pine Creek Valley and an equal extent of the Queer Creek Valley, reaching almost to South Bloomingville. The deep layer of alluvium left by the now vanished lake in both valleys is responsible for the flat, rich farmland found there. The flooding did not affect the upper reaches of the streams, however, and they continued in their ancient courses. The rising water eventually found an outlet through a low place in the surrounding hills and the draining glacial lake quickly sliced a gorge into the bedrock which is now utilized by Salt Fork, the stream into which Pine Creek and Queer Creek presently flow. This most recent history is readily visible along State Route 56 between Laurelville and South Bloomingville where the flat floor of the wide valley represents the surface of the fill left in the bottom of the old glacial lake bed. A sign for Narrows Road points the way to the gorge cut through the bedrock hills by the raging waters released from the old lake. At South Bloomingville the traveler enters the Hocking Hills and the ancient headwaters of Queer Creek and Pine Creek, deepened by the events of the Quaternary Period.

Balcony Natural Bridge (OH-A-HOC-02) — Hocking County
Span: 11 feet (3.3 meters); Clearance: 17.5 feet (5.5 meters)
Balcony (formerly Fosters) Natural Bridge (Figure 89; Plate 7) is located in Hocking State Forest at the head of a small hollow which opens into The Gulf of Queer Creek. The bridge was named for a distinctive sandstone ledge jutting out beneath it. Although a small arch, it has a dramatic location 18 feet above the floor of the narrow valley. The lintel is 11 feet long and 3 feet wide at its narrowest point. It averages 1 foot in thickness, but is only half that at the thinnest place.

Figure 89. Balcony Natural Bridge. The lower ledge of rock is the "balcony" for which this arch is named.

175

As with many of the Hocking arches, Balcony Natural Bridge was formed when the roof of an alcove was breached. An obvious joint plane angling to the northeast has played an important role in shaping the bridge. It forms the face of the cliff just below the western end of the lintel and can be traced across the arch itself. Water from the small stream which formed the alcove must have found its way down into this joint, enlarging it into the skylight behind the bridge. The bridge will eventually fail along this same joint since it forms a line of weakness across its western abutment.

An interesting domed recess is found in the ceiling of the alcove remnant west of the bridge. It is most likely the result of sapping, a process in which weathering weakens rock and causes the individual grains composing it to fall away. The same process may also have been involved in enlarging the skylight of the arch. The alcove beneath the arch extends considerably farther on either side than does the skylight. It is a double-decked alcove with a protruding ledge of more-resistant sandstone along its back wall, forming the balcony for which the arch is named. Such multi-story alcoves are not uncommon in the Hocking Hills. While the Black Hand Sandstone usually exhibits the simple sandwich configuration of a weak layer supported and capped by harder layers, variations do exist. These more complex "sandwiches" can produce stacked alcoves when weathered, the number of alcoves depending on the number of weak layers available to be hollowed out. The cluster of alcoves and other features at Balcony Natural Bridge have all been carved out of the upper hard layer of the Black Hand sandwich. This fact is emphasized by the stream which falls through the arch and then flows away at a slight grade as it cuts down into the soft middle layer of sandstone. Reaching the top of the lower hard layer of the sandwich, it falls into another small gorge, the site of Three Hole Arch.

Three Hole Arch (OH-A-HOC-17) — Hocking County
Span: 3.5 feet (1.1 meters); Clearance: 1.8 feet (0.5 meter)

Three Hole Arch opens into a low cliff on the east side of the waterfall at the head of the lower box canyon below Balcony Natural Bridge (Figure 90). Although the waterfall is less than 20 feet high, its slick face combined with the lack of adequate handholds make scaling it a

176

Figure 90. Three Hole Arch. This is a pillared-alcove arch with three entrances.

foolhardy enterprise and should not be attempted. Three Hole Arch is not large, but this pillared-alcove-type arch is unusual in having three entrances, although only two are visible from below the falls. The eastern entrance is 3.5 feet wide and 3.4 feet high. The middle and largest entrance is 6.5 feet wide and 2.5 feet high, leaving a pillar 5.4 feet wide between them. The third entrance is nearly circular in outline with a length of 2.25 feet and height of 1.4 feet with a 4 foot wide pillar separating it from the middle entrance. The opening runs in a very gentle curve from the eastern entrance to the western, circular entrance. The middle entrance opens through the outside wall of the arch; in this it is similar to, but much smaller and less complicated than, Rock House. The span of 3.5 feet and clearance of 1.8 feet were measured at the smallest constriction of this opening which is behind the pillar between the eastern and middle entrances. Although the east entrance is slightly lower on the cliff face than the other two, all three entrances are basically in line, and they and the opening behind them have obviously been eroded into a horizontal layer of less-resistant rock.

The sand-covered floor of the arch is fairly level until it reaches the circular entrance, which it slopes down slightly to meet. This slope, along with the outline of the entrance, brought to mind an appropriate name for it: "the drain." But is it a drain in fact as well as shape? Here is a good place to contemplate an important question relating to all the similar pillared-alcove arches which are found in the area: Where did all the sand that once filled the opening go and how was it removed? Once the cement holding the sandstone together has been dissolved by groundwater or atmospheric erosion, the released sand grains are free to fall away, but they need a void to fall into. Near the entrances they can simply fall out into the adjoining gorge, but once the angle of repose of loose sand has been reached, gravity will no longer remove the sediment. Wind might possibly contribute to the removal process, but given the constricted nature of the gorge at this point and the small size of the entrances, it is doubtful that any but the strongest gusts would be able to penetrate even this shallow opening to any degree. Rain might remove some of the sand, but it could only do so at the entrances. Even at that, the surface area of loose sand is so great compared to the amount of rain that could reach it that most of the water entering the arch by this means would immediately be adsorbed onto the surface of the sand grains and locked up, eventually evaporating without doing any mechanical work at all. Animals seeking shelter might be partly responsible for a limited amount of sand removal, especially if they were also foraging for food, but this particular cavity is not easily accessible and, to my eyes at least, provides little to attract a hungry creature.

It might be possible that such pillared-alcove arches were cut by running water when the streams now running below them were flowing at the same level as the openings. However, while alcoves and recesses are presently found in great numbers alongside streams, at no place have any pillared alcoves been observed in circumstances that indicate that a stream was responsible for forming them.

At Rock House and some other pillared-alcove-type arches, seepage along horizontal bedding planes at the sides and rear of the arch appear to provide enough water flow to slowly move loose sand toward the entrances where gravity and wind can do their work. The flow at

Rock House is especially strong and it is easy to see how such a process could operate. Here at Three Hole Arch, however, there is no noticeable seepage coming out of the walls and the sandy floor is perfectly dry.

And yet this dryness may be somewhat deceiving. A narrow film of water, barely enough to wet the rock, covers the face of the low cliff below the east entrance. It may be that water seeps from the interior walls of the arch and moves across the floor beneath the sand covering to escape out the entrance. Such seepage may be strong enough at times to slowly move, or at least lubricate, the sand, facilitating its movement out of the arch. A cone of loose debris below the east entrance would seem to verify that this movement actually does take place. Such are the cogitations of an arch explorer while sitting in a quiet alcove beside a soothing waterfall after a long, hard day.

A short distance downstream from Three Hole Arch stands the Gateway — two large bedrock masses facing each other on opposite sides of the stream (Figure 91). They almost appear to be the abutments of a now-fallen natural bridge of impressive size, but are more

Figure 91. The Gateway leading to Three Hole Arch as viewed from above.

likely just an expression of more-resistant knots of sandstone brought into relief by incessant weathering.

Brineinger Hollow Natural Bridge (OH-A-HOC-03)
— Hocking County
Span: 16.75 feet (5.1 meters); Clearance: 3.3 feet (1 meter)

Located on the north wall of Brineinger Hollow, Brineinger Hollow Natural Bridge (Figure 92) was carved by an intermittent stream which has so far only managed to cut a groove into the steep talus slope beneath the high cliff lining the valley. The rock forming the bridge is part of the bottom layer of the Black Hand Sandstone, showing that in places each of the three layers are themselves composed of layers of varying hardness. The stream at this point has sapped out a small alcove 15 feet deep with a reach of 31 feet. A crevice somewhat parallel to the front of the alcove has been enlarged into a narrow skylight 16.75 feet long and 4 to 6 feet wide. The remaining front rim of the recess which forms the bridge is 11 inches thick at the back, but narrows to half that at the front.

Figure 92. Brineinger Hollow Natural Bridge. Standing on the bridge, Jeff Johnson, District Manager with Ohio DNAP, provides scale.

The back side of the lintel exhibits the smooth, vertical face of a joint plane, but simple crevice enlargement has not been the only process responsible for forming the skylight. There is much evidence of ceiling collapse in the form of rock slabs 4 to 5 feet wide and of the same thickness as the lintel.

Brineinger Hollow is now part of Crane Hollow State Nature Preserve which is owned by Crane Hollow, Inc. Permission of the owners is required to visit it.

Chapel Ridge Natural Bridge (OH-A-HOC-04)
— Hocking County
Span: 12 feet (3.7 meters); Clearance: 4 feet (1.2 meters)

Chapel Ridge Natural Bridge is unusual for Ohio in that its skylight is a pothole. In the Colorado Plateau, pothole arches are not uncommon. They form when a deepening alcove near the rim of a sandstone cliff meets a pothole enlarging downward from the upland surface. In this case, the pothole is a result of precipitation collecting in a slight depression and dissolving the cement holding the grains of the sandstone together. When the water dries up, the sand blows out of the depression and the pothole is deepened ever so slightly. Over time these pits can become quite large. The pothole of Ohio's pothole arch formed in a different way. In this case, both the alcove and the pothole were created by a small stream tumbling over the back wall of a box canyon. Stream potholes such as this one can form at and near falls and rapids where strong, revolving currents swirl sand and gravel against the bed of the stream, grinding down into the rock like a slow-motion drill. Round or oblong steep-sided pits, sometimes large enough for a human to lie in, are the result. A fine and easily viewed example is Devils Bathtub (Figure 93) just downstream from the Upper Falls at Old Mans Cave.

The Chapel Ridge pothole is found in the middle level of a typical Hocking Hills multi-level waterfall. It formed to the side of the stream responsible for it, perhaps as a result of the rushing water finding a small mass of more weakly cemented and easily eroded rock than that in its bed. The resulting pothole is 12 feet wide and 8 feet deep. At this place, the middle layer of the Black Hand Sandstone sandwich is itself formed of several thinner layers of varying hardness. A

Figure 93. Devils Bathtub, a stream-carved pothole in the bed of Old Mans Creek at Old Mans Cave in Hocking Hills State Park.

layer of harder rock capping a layer of softer rock has allowed the formation of a minor waterfall with a 4 foot high alcove behind it. As this alcove deepened, it eventually broke into the lower part of the pothole, capturing some of the stream's flow and creating the natural bridge.

The shape of the pothole indicates that it was cut to its present depth before the enlarging alcove reached it; that is, the pothole was breached by the alcove. The remaining walls of the pothole are vertical and curve into an oval outline through its entire depth, showing that they were cut by swirling water. As soon as the front wall of the pothole was broken by the enlarging alcove and the water had a direct exit, its swirling action stopped and enlargement of the pothole ceased.

The 12-foot length of the pothole defines the length of the lintel, although the alcove widens rapidly away from it toward the cliff face. The lintel averages 5 feet in thickness.

Conkles Hollow Arch (OH-A-HOC-05) — Hocking County
Span: 10 feet (3 meters); Clearance: 8 feet (2.4 meters)

In keeping with Appalachian usage, narrow, steep-sided valleys in southeastern Ohio are often called "hollows." Conkles Hollow in the Hocking Hills region is considered to be the deepest such gorge-like valley in the state with cliffs that rise in places over 200 feet. It was named for an inscription found on a rock many years ago that recorded the name of W. J. Conkle and the date 1797.

In the mid-1980s, a more recent member of the Conkle family still living in the area informed Mark Howes, manager of Conkles Hollow State Nature Preserve, that a natural arch existed on the northern wall of the hollow. Although later research showed that a record of this feature had been filed with Ohio DNAP in 1983, it was not investigated until Mr. Conkle's mention of it prompted renewed interest. It was the rediscovery of this arch — Conkles Hollow Arch (Figure 94) — which inspired the Ohio Natural Arch Survey. Its location unfortunately does not permit public access.

Conkles Hollow Arch cuts through a large, rounded buttress-like apron projecting from the wall of Conkles Hollow. The apron owes its existence to stronger cementation of the sandstone at this location. The core of the apron is more weakly cemented, however,

Figure 94. Conkles Hollow Arch. This is the arch that launched a book.

and removal of this sandstone by weathering has produced the arch opening which has a height of 4.2 feet and a span varying from 10 feet at the south end to 16 feet at the north. The exterior surface of the apron exhibits case hardening, which is a coating of cementing material carried in solution by moisture moving outward through porous rock and deposited at the surface when the moisture evaporates. This protects the rock and slows its disintegration. The walls within the shelter of the arch lack this protection, however, and are actively sloughing off, displaying one method by which the arch has been, and continues to be, enlarged. Cross-bedding is quite noticeable in the apron and no doubt helped to define the shape of the arch opening. The floor is level and extends beyond both entrances of the arch, indicating that it represents the top of a more-resistant layer of sandstone.

Polypody Arch (OH-A-HOC-06) — Hocking County
Span: 8.75 feet (2.7 meters); Clearance: 3.7 feet (1.1 meters)

In 1993, Mark Howes, manager of Conkles Hollow State Nature Preserve, reported sighting a possible second arch within the preserve. Its presence was later confirmed by other employees of Ohio DNAP. It was named Polypody Arch (Figure 95) after the Common Polypody

Figure 95. Polypody Arch. Note the three men sitting in the opening of the arch.

(*Polypodium vulgare*) (Figure 96), a species of fern which is often found growing in the sheltered hollows of the Hocking Hills.

Like Conkles Hollow Arch located on the opposite side of the gorge, Polypody Arch is a pierced apron of sandstone extending from the cliff. The cliff face forms a side wall of the arch, indicating that a joint plane may be involved. This graceful arch has a clearance of 3.7 feet measured at its lowest point which is about halfway through the arch. The height of the opening increases to 5.75 feet at both entrances, reflecting the work of atmospheric weathering on these

Figure 96. Polypody Arch. The opening of this arch passes behind the rock abutment. The fern in the foreground is Common Polypody (*Polypodium vulgare*), the plant after which this arch was named.

Figure 97. *Sullivantia sullivantii,* a rare endemic plant of the American Midwest.

more exposed faces of the arch. The 8.75-foot span remains fairly constant throughout, widening only slightly at the entrances. A vertical crevice which forms an eroded offset can be followed up the north wall of the opening and then diagonally across the roof toward the south wall. However, it disappears before reaching the wall, showing that crevices in the Black Hand Sandstone can be discontinuous.

The small valley containing Polypody Arch also contains several examples of alcoves and small waterfall sites where intermittent streams cascade down the steep walls — erosional forms typical of the Black Hand Sandstone. The area is also notable for a healthy population of *Sullivantia sullivantii* (Figure 97), a rare plant only found in the American Midwest.

Hagley Hollow Arch (OH-A-HOC-07) — Hocking County
Span: 26.5 feet (8.1 meters); Clearance: 6.2 feet (1.9 meters)
Hagley Hollow Arch (Figure 98; Plate 8) resembles the two Conkles Hollow arches in that its opening is pierced through an apronlike buttress extending out from the wall of a hollow. The opening is

Figure 98. Hagley Hollow Arch. This view through the arch opening looks toward the cliff at the head of Hagley Hollow.

8.33 feet high at the upstream entrance and 7 feet high at the downstream entrance. The 26.5-foot span of this arch equals its width. Some cross-bedding is visible at the south end, but seems to have had little effect on the shape of the arch. Most of the visible layering at this location is nearly horizontal.

The initial report of this arch came by way of a letter to Ohio DGS from Robert Vreeland who found it as part of his continuing research for a series of guidebooks on the natural arches of the United States. He recorded its span as 31 feet and its height (clearance) as 7 feet, measurements which differ slightly from those given in this report. Here is a good example of two experienced arch measurers arriving at different figures because they chose slightly different parts of a given arch to measure.

Unger Hollow Natural Bridge (OH-A-HOC-08)
— Hocking County
Span: 24 feet (7.3 meters); Clearance: 6 feet (1.8 meters)

Unger Hollow Natural Bridge (Figure 99), named for the hollow in which it is located, is a breached-alcove arch, but with significant variations from the norm. The skylight is located at the rear of the alcove and to one side, giving the arch a complicated, lop-sided floor plan. The alcove itself has a reach of 53 feet, but a height of only 6 feet which diminishes slightly toward the back wall, creating a very wide, low space. The skylight opens into an extension of the alcove at its southeastern corner. The intermittent stream responsible for this work falls through the skylight and flows in a shallow trough which follows the back wall of the alcove in a wide loop to the entrance, although gravel and silt on the floor show that when the stream floods, it can fill the entire alcove.

The skylight has a length of 24 feet and this is given as the span of the natural bridge. Much of the complexity of this feature is due to its skylight. It opens through the floor of a small alcove cut by the stream into the side of a hill above the larger alcove. Before this minor recess was deepened enough to breach the roof of the larger

Figure 99. Unger Hollow Natural Bridge. Water falling through the skylight at the back of the alcove is visible near the center of the picture.

alcove below, the stream appears to have had a course across the roof of the lower alcove as shown by a very shallow rock-cut trough the same width as the stream valley above the natural bridge. After the breach was made, the stream dropped through the now-open skylight, abandoning the short section above the lower alcove in favor of its present course across the alcove floor. The ceiling of the upper alcove forms a "roof" which juts out over the skylight, blocking the view of the sky from beneath it except along a short stretch at its upper end.

Presumably the main alcove was cut first, since there had to be an alcove to breach. The stream may have enlarged and deepened this recess, but the alcove's location to the side of the stream's original course would seem to indicate that the stream was not the main agent of weathering. Seeps in the wall of the alcove just above floor level may have played a larger role. The majority of these seeps are found in the extension of the lower alcove beneath the skylight. Some of the strongest of them are near the falls where the stream plunges through the skylight and almost certainly are fed or at least augmented by water from the stream seeping down through crevices in the bedrock.

However it formed, the alcove must have been nearly as large at the time of breaching as it is now since the skylight is located above its deepest penetration into the hillside. The abandoned course of the stream indicates that the valley below the alcove had its present shape and size before the alcove was breached. Compared to the narrow, rock-choked valley above the alcove, the lower valley is wide and level. The difference is striking and is probably due to two horizontal layers of rock with differing resistances to erosion, the upper one being harder than the lower.

Unger Hollow Natural Bridge is a good example of a feature meeting the definition of "natural bridge" without actually looking like one. Rather than an arch of rock, it appears more like a hole in the ground that just happens to connect to an alcove. It is, however, a true natural bridge and a rather perplexing one at that.

Old Mans Pantry (OH-A-HOC-09)* — Hocking County
Span: 20 feet (6.1 meters); Clearance: 3.5 feet (1.1 meters)
Old Mans Pantry (Figure 100), an opening in the cliff along Old Mans Creek, is one of the most easily viewed pillared-alcove-type

Figure 100. Old Mans Pantry. The two entrances to this pillared-alcove-type arch as seen from the trail along Old Mans Creek in Hocking Hills State Park.

arches in the state. Although not uncommon in the Hocking Hills region, most pillared-alcove arches are located deep within wild hollows and require some effort to see. Old Mans Pantry, on the other hand, is readily visible from the trail leading downstream from the Upper Falls at Old Mans Cave. While showing obvious signs of frequent visitation by people wandering off the legal trail, it has not attracted enough attention to receive a name. The one given it in this report is evidently the first recorded for it.

The name seems appropriate because this arch takes the form of a room opening up behind the cliff face, complete with a "doorway" and "window." The larger of the two openings is 8 feet wide and 9 feet high and comes equipped with a 1-foot-high step. Inside this doorway, a narrow ramp slopes steeply up to the flat, main level of the arch. The ceiling slopes down to meet the floor at the back. An oblong window 13 feet long looks out over the stream. The blocky pillar separating the door and the window is 3 feet wide and 7 feet thick.

With 20 feet between the pillar and the back wall, Old Mans Pantry would provide adequate dry storage space for the old man's food supplies, although they would be none too secure. The "old man" of Old Mans Cave was a fugitive from West Virginia named Rowe who set up housekeeping sometime after the Civil War in the large alcove, now called Old Mans Cave, downstream from the pantry. When his hermit life came to an end, he was buried beneath the rocks of his wild home. Although called a cave, Old Mans Cave is really an alcove eroded into the weak middle layer of the Black Hand Sandstone, as are most of the so-called "caves" in the Hocking Hills region.

Old Mans Pantry, on the other hand, has been weathered into the resistant upper layer of the Black Hand Sandstone, the same layer which forms the Upper Falls a short distance upstream. The horizontal elongation of the window of the pantry coupled with the presence of the doorway and a few other very small openings along the same line indicate that weathering here has opened an alcove along a bedding plane within this upper layer. Although no moisture was evident when the site was visited, seepage has no doubt been an important factor in its formation. It is possible that the "door" opening of the pantry has been enlarged with human help, although no tool marks or other evidence of such alteration is visible.

Rock House (OH-A-HOC-10)* — Hocking County
Span: 20 feet (6.1 meters); Clearance: 40 feet (12.2 meters); Tunnel Length: 185 feet (56 meters)

Rock House (figures 101–104; plate 9 and 10) is unique in Ohio. It is by far the largest known natural arch in the state, one of only two natural tunnels surveyed to date, and the only one with "Gothic windows" opening through its side. This geological treasure has attracted the interest of tourists and scientists alike for many years and its popularity led to it becoming one of the first units of Hocking Hills State Park to be purchased.

This impressive arch has been carved from a 115-foot-high cliff rising above a small tributary of Laurel Run. At least two major vertical joints parallel this small stream. One forms the main cliff face; the other, located closer to the stream, marks the front wall of Rock House. Much of the rock between these two joints has been removed except for the

Figure 101. Rock House. A view through the west entrance from inside the arch.

massive block into which Rock House has been carved. It remains as a remnant jutting out from the valley wall (Frontispiece and Figure 102).

The joint forming the main face of the cliff continues behind this large, rectangular block and has been enlarged to form a tunnel which is over 200 feet long. The passage is open to the sky at either end of the tunnel and these sections are not included in the span length. The tunnel is approximately 40 feet high and 20 feet wide at floor level, although this last figure is quite variable. Several joints cross the main one at varying angles and can be traced from the back wall across the floor and out the front wall of Rock House where they have been widened to form its "windows."

Weathering of the weak middle layer of the Black Hand Sandstone along these joints has given us this natural tunnel. Most of the

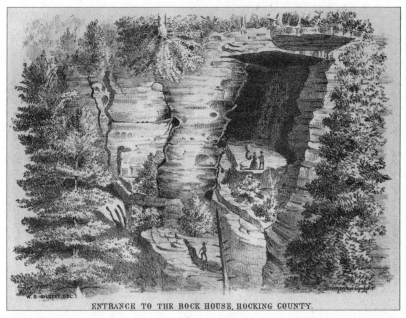

ENTRANCE TO THE ROCK HOUSE, HOCKING COUNTY.

Figure 102. Rock House. Drawing of the exterior of the west entrance from the Geological Survey of Ohio's Report of Progress in 1870. This entrance marks the west end of the main vertical crevice forming the arch. The right wall of the crevice continues the line of the main cliff. The arch opening has been eroded through a remnant block of sandstone forming a massive buttress against the cliff.

joints carry water, especially after heavy rains. This near-constant moisture against their walls has dissolved the cementing material holding the sandstone together, allowing the grains of sand to drop away. The loose sand is then carried out through the openings. Wind may play a small role in this process, but the steady flow of water has been far more important. The present rate of seepage might appear to be too weak to accomplish much transport, but over time it would be quite adequate. It is also possible that the flow was augmented during Pleistocene time when the nearby ice sheet could have encouraged increased precipitation.

For those who still think that the size of Rock House demands a more dramatic origin than mere seepage and solution along rock fractures, Carman (1946) tells us that, if Rock House were cut entirely during the Pleistocene, removal of the 100,000 cubic feet of sand that once filled it would require that only one-half ounce of sand a day be carried out. That amount is approximately equal to two level table-spoons. If Rock House began forming before the start of the Pleis-tocene, then the amount of sand required to be removed daily would be even less.

This tunnel is a striking example of the control of form by joints (Figure 103). Each of the five windows opening through its side wall is actually one of the perpendicular joints which has been enlarged. These joints can be followed from the windows across the floor of the tunnel and up the opposite wall where alcoves have been hol-lowed out by groundwater seeping through the crevices. These al-coves have ceilings which slope down from where they open onto the main passage of the tunnel to the back of the alcove where they meet

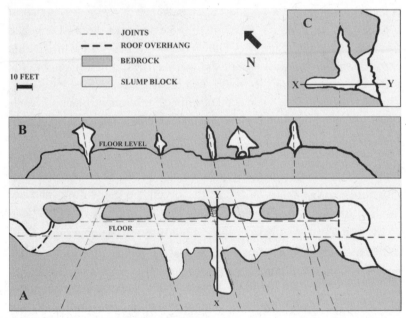

Figure 103. Rock House. **A:** The floor plan of the arch. **B:** A diagram of the outer wall from inside the arch showing the shapes of the "Gothic-arched" windows. **C:** A vertical cross section of the arch from the back wall to beyond the main entrance as measured along the section X-Y as marked in part A. (Adapted from Carmen, 1972)

flat floors. While some are small, several are quite respectable in size. One of the largest has an opening which is 13 feet wide and 4 feet high and extends more than 20 feet into the rock. The floors of many of these alcoves lie at the same level as, or a few feet above, the floor of the main tunnel passage (Figure 104). They mark the top of a more-resistant layer of sandstone which also forms a projecting bench along the wall of the tunnel.

Enlargement of the joints most likely proceeded simultaneously. The seeps now found in the back wall would have started at the cliff face and eroded their way back into the hillside. At the same time, water moving down into the main joint from above and exiting at

Figure 104. Rock House. A photograph of the interior looking down the length of the main enlarged vertical crevice (Van Tassel, 1901). The "Gothic-arched" windows are on the right. The view has changed little since this picture was taken around 1900.

either end would have enlarged it to form the main passage. The widening crevices eventually coalesced into the feature as we have it today. The loose sand covering its floor shows that such enlargement is still continuing, undoubtedly helped by the abrasive impact of thousands of tourist feet. The "Gothic" profiles of the main opening and the windows — narrow at the top and widening toward the bottom — are typical of such enlarged joints. Their shapes have also been influenced by cross-bedding which is evident in the sandstone of the walls.

While iron compounds supply the cement holding the Black Hand Sandstone together in much of the Hocking Hills, the more soluble calcium carbonate appears to be a major cementing agent at Rock House as indicated by thin, botryoidal encrustations on some of the sheltered rock walls within the tunnel. This cement is more easily removed than iron compounds, and may have been one reason for the formation of Rock House at this particular location. These calcium carbonate encrustations contribute to the colorful walls of the tunnel, a feature noted by many who have written of it. Various oranges, reds, browns and whites mottle the surface of the sandstone, forming a pleasing contrast to the greens and grays of moss and lichen.

Interest in Rock House goes back many years. Indians were the first to use it. Later, if tradition can be trusted, horse-thieves, robbers, murderers and bootleggers found it a convenient haven. Tourists came early and stayed long. In 1835, a small hotel was built to accommodate them where the state park picnic shelter now stands. For a short period of time Rock House even had its own post office. There is no record as to how the tourists got along with the robbers, murderers and horse-thieves. Presumably relations were quite jolly with the bootleggers.

Annex Arch (OH-A-HOC-18) — Hocking County
Span: 5.6 feet (1.7 meters); Clearance: 2.6 feet (0.8 meter)
Annex Arch (Figure 105) was named for its location less than 100 feet from Rock House. Although considerably smaller than its large neighbor, Annex Arch shares with it its manner of formation. Annex Arch is a bedrock-texture-remnant arch formed by atmospheric weathering of rock, with varying resistance to erosion, within the Black

Figure 105. Annex Arch. Cross-bedding is exhibited in the massive column where the angled beds in the lower, lighter part contrast sharply with the horizontal beds in the upper part.

Hand Sandstone. The opening pierces a low fin of rock left behind when less-resistant rock was weathered away on either side. The remnant of the fin forming the outer abutment is only 5 feet long and 3.3 feet wide, but it has a massive appearance nonetheless and seems well suited for holding up the thick, more-resistant layer of sandstone above it. This massive layer forms a canopy that juts out several feet beyond the outer abutment and is one of the striking features of this arch.

M Arch (OH-A-HOC-11) — Hocking County
  Span: 18 feet (5.5 meters); Clearance: 4 feet (1.2 meters)
    M Arch (Figure 106), named for the appearance of contour lines on a US Geological Survey topographic map of its location, has been opened through a monolithic sliver of Black Hand Sandstone 32 feet thick separated from the adjoining cliff by an enlarged crevice less than 3 feet wide. A vertical joint running through the sandstone defines the south side of the arch, appearing as a comparatively flat wall. From here the roof of the arch extends north in a sloping curve,

Figure 106. M Arch. This view shows the opening of the arch as seen from the widened crevice which gives access to it.

creating what would be a near-perfect half-arch shape were it not for an irregular knob of bedrock that intrudes on the space. This narrows the opening considerably at its center, although it widens to a respectful height at each end. The clearance was measured at the southern end where the opening is the highest.

The lintel of this arch is very massive, being 17 feet thick at the south end, but gradually increasing to a thickness of 21.5 feet at the north end due to the downward slope of the arch's ceiling. The lintel is 25 feet wide, part of the outer face of the monolithic sliver having evidently been lost at this point. The only access to this arch opening is from the enlarged crevice; on the outer side of the sliver there is a sheer drop of 47 feet from the base of the arch to the ground below.

The opening of the arch pierces the rock sliver, which is several hundred feet long, within 23 feet of its northern end. How it formed is something of a mystery. The quarter-circle shape might indicate that rock fracturing and tension dome mechanics may have had a part to play, but the persistent knob of bedrock nearly filling the opening would seem to rule that out. The best explanation would appear to be that it formed through differential weathering of a less-resistant section of rock, the interior knob being more resistant. If this is so, it is interesting to note that no evidence of softer rock such as a shallow alcove is seen in the wall of the enlarged crevice opposite the opening. Either the softer rock ended at the crevice, or it extended only part way through what is now the opening of the arch and rock fracturing and collapse took care of the rest. There is not much rock debris on the floor of the arch, but the steep drop at its outer entrance and the downward-sloping floor of the enlarged crevice onto which its interior entrance opens would provide adequate routes for removal, especially after the rock fragments had broken down into loose sand.

This section of cliff wall boasts several interesting features. Balanced Rock and Flat Iron Rock, discussed in the section on Ohio's natural pillars, are found at the opposite end of the monolithic sliver. M Arch is located within privately owned Crane Hollow State Nature Preserve and permission from the owners must be obtained before visiting it.

1811 Arch (OH-A-HOC-12) — Hocking County
Span: 40 feet (12.2 meters); Clearance: 16 feet (4.9 meters)

1811 Arch (Figure 107) gets its name from the earliest date found carved into its walls. In 1811, Ohio had been a state for only eight years, Indiana was still a territory, the British still held Fort Detroit and Tecumseh was getting ready to go on the warpath. Although such graffiti is rightly considered a blemish, especially the much larger and gaudier modern spray-painted versions, in this case it brings the distant past close to hand. It also shows that, although this arch is not large or showy, it has been frequently visited over time, most likely a result of its location near one of the major roads of the region.

Weathered into the lower level of the Black Hand Sandstone, 1811 Arch is a pillared alcove 52 feet deep with two entrances opening at right angles to each other. The larger entrance is 40 feet wide

Figure 107. 1811 Arch. The massive column dividing the entrance creates a pillared-alcove-type arch.

and 15 feet high, the smaller one is 17 feet wide and 4 feet high. They are separated by a remnant of the cliff face 13 feet wide.

Early Arch (OH-A-HOC-13) — Hocking County
Span: 30 feet (9.1 meters); Clearance: 6.7 feet (2 meters)

Early Arch (Figure 108), found on the upper reaches of a small stream, is a large pillared alcove with two entrances opening side-by-side high on the cliff face. Its origin as two enlarging alcoves which eventually united is fairly obvious. The resulting pillar left between the entrances is 8 feet wide and 12 feet thick. The east entrance is the larger of the two, being 47 feet wide and 11.5 feet high. The smaller west entrance is 36 feet wide and 10.5 feet high.

A nearly circular dome formed by the collapse of a 26-inch-thick slab of sandstone can be seen in the roof of the eastern alcove (Figure 19). This is one of the best examples of a tension dome ceiling found in the state.

Early Hollow, in which the arch is found, was named for a local family. Much of the hollow is owned by Crane Hollow Inc. Permission of the owners must be obtained before visiting the arch.

Figure 108. Early Arch. This is a breeched-alcove arch formed by the uniting of adjacent alcoves.

Saltpetre Cave Arch (OH-A-HOC-14)* — Hocking County
Span: 114.2 feet (34.8 meters); Clearance: 9 feet (2.7 meters)

In 1835, a powder mill was built near the town of Gibisonville to take advantage of the saltpeter deposits found in the recesses of the nearby hollows. Saltpeter, or potassium nitrate, was gathered as "nitrous earth" from protected alcoves and used to make gunpowder, a valuable commodity in what was then a wild part of Hocking County. Almost the only relic of this old industry is the name given to Saltpetre Cave Arch (Figure 109), one of the alcoves from which saltpeter was apparently removed, located just 3 miles from Gibisonville. This history, along with the romantic spelling of the cave's name, became part of a short-lived commercial tourist venture in the 1990s. The signs announcing it were probably the first indication most visitors to the Hocking Hills had that such a place even existed. After the demise of Saltpetre Cave Arch as a tourist destination, the cave and adjacent land was sold to Ohio DNAP and is now Saltpetre Cave State Nature Preserve. A permit is required to visit this area.

The "cave" is actually a pillared-alcove-type natural arch. In relation to its width, the alcove is one of the deepest found in the

Figure 109. Saltpetre Cave Arch. The two entrances as seen from inside this pillared-alcove-type arch.

region, extending 126 feet into the hillside and having a reach at the entrance of 45 feet. The western entrance has a height of 9.75 feet and width of 18.75 feet. The smaller east entrance is 7.5 feet high and 11.8 feet wide. The pillar between them is 11.5 feet wide at the base and 12.5 feet thick, giving it a very blocky presence.

Although this arch may have started as two alcoves eroding into the cliff face on either side of the pillar, they quickly merged. There is no evidence beyond the pillar of a common wall having been breached as is usually found where two alcoves eventually join through enlargement. The "cave" is essentially a single alcove. There is very little rock debris on the sandy floor, but thin slabs of sandstone less than an inch thick detaching from the ceiling show that the opening continues to enlarge through rock collapse, although at a very slow rate. Whitish blooms of potassium nitrate still spot the ceiling and walls. These mineral salts are dissolved out of the bedrock by groundwater and carried to the surface where they are deposited when the water evaporates. It would seem a tedious job indeed to scrape enough of this thinly crusted mineral to make the gunpowder business profitable.

That may be one reason why the industry eventually died out, although an unfortunate explosion which destroyed the mill at Gibisonville was certainly another factor.

Surprise Arch (OH-A-HOC-15)* — Hocking County
Span: 15.5 feet (4.7 meters); Clearance: 1.9 feet (0.6 meter)

Unlike its near neighbor Saltpetre Cave Arch, Surprise Arch (Figure 110), is most definitely a result of the junction of two adjacent alcoves. The opening through the common wall of the alcoves is long and narrow and not easily seen. In fact, this arch was not known to exist until Jeff Johnson, District Manager with Ohio DNAP which now owns it, happened to notice a glimmer of light where none was expected. Closer inspection revealed the nearly hidden opening of the arch.

The two alcoves are impressively large and getting larger as shown by huge slabs of collapsed roof rock which broke apart on landing, the pieces remaining in position like a very thick jigsaw puzzle. The breakdown in the west alcove comes with a story popularized by tour guides during the commercial days of Saltpetre Cave.

Figure 110. Surprise Arch. This narrow opening joins two alcoves.

During the frontier era when Euro-American settlers were pushing their way into Indian country west of the Appalachian Mountains, a band of Shawnees met in this alcove to plan an attack on emigrant boats floating down the Ohio River. During their meeting, the heat from their council fire ignited the saltpeter (potassium nitrate) incrusting the ceiling of the alcove. The resulting explosion brought down the great slab of rock now seen there, burying forever the entire band of Indians.

Aside from the fact the potassium nitrate is not readily flammable and certainly not explosive until mixed with sulfur and charcoal to make gunpowder, the story falters on two other counts. First, American Indians did not build large fires as a rule, preferring to use only enough fuel to provide for heating and cooking. It was the white pioneers who insisted on blazing bonfires, a habit that Indians often derided. Second, if Indian carelessness explains the large rock fall in the western alcove, what explains the equally large rock fall in the eastern alcove? The Shawnee may have been outgunned, but they were rarely outsmarted. Having suffered one such catastrophe, they would not have courted disaster a second time by holding another conference complete with roaring fire next door to where such a memorable event had occurred. Still, the story remains an interesting example of a popular tale that shows how far guides could stretch history for a paying audience.

Muddy Crack Arch (OH-A-HOC-16) — Hocking County
Span: 1.2 feet (0.4 meter) Clearance: 8.6 feet (2.6 meters)
The measurements given for the span and clearance of Muddy Crack Arch (Figure 111) are a bit deceptive, for the arch itself as defined by the skylight is just one small part of a much larger feature. Muddy Crack is an enlarged vertical crevice that extends nearly 100 feet into the rock. Unlike many similar enlarged fissures in the Hocking Hills, this one is roofed through its entire length except for the small skylight. Its main opening, triangular in shape, has a respectable reach of 23 feet and a height of 12 feet at the entrance. The width of the opening is due to the loss of rock between the main crevice and a second vertical crevice roughly paralleling it. The section thus enlarged penetrates the hillside for a distance of 40 feet. A 6-foot offset,

Figure 111. Muddy Crack Arch. This view of the entrance shows the enlarged vertical crevice forming the arch which is defined by a skylight opening through the top of the crevice.

its flat face indicating another joint running at right angles to the first two, narrows the passage considerably after that. From this point on, the opening consists of just the enlarged main crevice which gradually narrows until, at 30 feet from the offset, it can no longer be traversed by ordinary humans. Although from this point on the crevice is only inches wide, it continues for approximately 30 more feet where its end is marked by a glow of light. After some searching, this second opening was discovered in the cliff face around the corner from the main opening at the base of a beautiful example of a blind arch — an arch-shaped alcove with a flat back wall. While large enough at the entrance to allow a slim human to squeeze in, the opening quickly narrows to impassibility.

The narrow skylight defining this arch was created by the enlarging vertical crevice breaking through to the surface. It is 5 feet long and only 1.2 feet wide, and pierces the ceiling just before the opening is narrowed by the offset. The narrow lintel thus formed is 13 feet wide from the end of the skylight to the edge of the cliff face above the main entrance. There is a drop of 23 feet from the deck of the lintel through the skylight to the floor of the crevice below. Of this, 14.4 feet is considered to be the thickness of the lintel since below that point the opening abruptly widens.

The name Muddy Crack comes from a pool of liquid, shoe-sucking mud that covers the entire floor of the arch beneath the skylight. Seen from above, the skylight has a definite funnel shape which helps to channel water into the crevice. Numerous drips from the ceiling of the crevice and the perimeter of the skylight show that this is indeed the source of most of the water. While dirt and debris almost certainly do make their way into the crevice through the skylight, most of the mud turns out to be wet sand, a result of the breakdown of the rock forming the sides of the crevice.

Arches in Other Mississippian Formations
Amherst Arch North (OH-A-LOR-01) — Lorain County
Span: 4 feet (1.2 meters); Clearance: 2.5 feet (0.8 meter)

Amherst Arch South (OH-A-LOR-02) — Lorain County
Span: 3 feet (0.9 meter) Clearance: 3.2 feet (1 meter)

The small Amherst arches (figures 112–114) are located on the north face of an outlier of Berea Sandstone (Figure 112). Each arch pierces a fin-like extension marking opposite ends of an 11-foot-high cliff capping the top of the outlier. Because the fins narrow toward their terminations, the widths of the arches through them also narrow. The width of the North Arch is 4.5 feet at the bluff end, but only 1.6 feet toward the end of the fin. The width of the South Arch is less variable, being 3 feet at the bluff end and 2 feet at the fin end.

Variation in stratification of the sandstone was an important factor in the formation of these arches. The outlier is topped by a 1-foot-thick cap of resistant sandstone. The rest of the bluff is made of thin-bedded sandstone showing some cross-bedding. There appears to be

Figure 112. The Amherst arches. An overall view of the abandoned head-land bluff that once was on the shore of glacial lake Whittlesey which cut the two arches by wave action. Amherst Arch North is readily visible to the left in the photograph.

less iron in the cement of the sandstone immediately beneath the cap rock as compared to the sandstone in the lower half of the outlier. The openings of the arches have formed in this weaker section of rock.

A hollowed-out alcove 4 feet wide extends 7 feet into the bluff between the arches and would have created a third arch were the bluff not considerably wider here. It is not difficult to imagine the same forces creating all three features, but what were those forces? Water-worn gravel similar to that seen on the Lake Erie shore a few miles to the north has been found nearby. This clue opens the possibility that the arches may have formed in a coastal environment when the lake was at a higher level. Lake Erie is just the most recent of a series of late-glacial-age lakes to occupy its present basin. As the last conti-nental glacier melted back to the north, its southern edge took on a lobed appearance due to less rapid wasting of the thicker ice found in the gouged-out basins that would become the Great Lakes of today. In Ohio, glacial meltwater was trapped between the ice front and the continental divide to the south, forming a long, crescent-shaped lake around the west end of the ice lobe occupying the Lake Erie Basin.

Figure 113. Amherst Arch North. This is a breeched-alcove arch formed in a now-vanished coastal environment.

This lake, called Lake Maumee, was the first hint of the system of great lakes that were to come.

As the ice front retreated farther, most of the Lake Erie Basin was uncovered and flooded by meltwater, forming Lake Whittlesey (Leverett and Taylor, 1915). This glacial lake was much larger than present Lake Erie, extending up the Maumee Valley nearly to Fort Wayne, Indiana, as shown by sand and gravel beaches that run across the landscape at an altitude of 735 feet above sea level, the same level as the surface of Lake Whittlesey. Further retreat of the ice opened up lower outlets for the meltwater to escape through, lowering Lake Whittlesey to the 675-foot level and creating Lake Warren. The last major pre-Erie glacial lake in the basin was Lake Lundy which left

Figure 114. Amherst Arch South. The opening of this arch has been eroded into thin-bedded layers of sandstone which are weaker than the massive cap rock.

several obscure ridges between 640 and 615 feet above sea level. The surface of modern Lake Erie stands at about 572 feet.

The shores of each of these glacial lakes is marked by beach ridges of sand and gravel, sandy delta deposits dropped by streams entering the ancient lakes and by typical water-cut coastal features where bedrock formed their shores. The Amherst arches, lying at 725 feet above sea level, were most likely formed near the shore of Lake Whittlesey which, like the other glacial precursors to Lake Erie, tended to fluctuate in depth during its brief existence. The main Whittlesey beach ridge, utilized by Middle Ridge Road, runs on a southwest-northeast line barely a mile to the southeast of the outlier containing the arches, which has the same orientation. During the existence of Lake Whittlesey the outlier no doubt formed an island just off the coast and later, as the water level lowered, a headland. The northwest side of the island faced into the lake and would have received the full force of its waves which could have carved out the arches and the alcove. This theory is further strengthened by the concave face of the bluff which contains the arches and by the presence of what is possibly lakeshore gravel nearby. Although now high and dry, the Amherst arches can be considered coastal arches due to the manner of their formation.

If true, this history gives these small arches an importance beyond their size. Such dramatic evidence of glacial lake erosion is not common. We are fortunate that they still exist. Berea Sandstone has been quarried for decades for use in constructing buildings. The area around the outlier did not escape. The bluff containing the arches sits between two abandoned quarries whose high walls tufted with vegetation rise from pools of dark water. Whether the bluff was preserved through intent or fortuitous accident will probably never be known, nor will we know what other erosional features were destroyed during quarrying operations. At this late date we can just be grateful that the arches have survived.

Raven Rock Arch (OH-A-SCI-01)* — Scioto County
Span: 15 feet (4.6 meters); Clearance: 7 feet (2.1 meters)

If natural arches in Ohio were rated according to the scenic quality of their site, Raven Rock Arch (figures 115–119; Plate 11) would certainly rank near the top. Located on the rim of a cliff of Mississippian sandstone, part of the Logan Formation, nearly 500 feet above the Ohio River, it opens onto a spectacular view of the storied valley.

The arch is just one of a number of interesting features found in this cliff. The site is named for Raven Rock, a flat slab of sandstone which projects 6.5 feet out into thin air from the top of the cliff. This rock is locally famous for stories of its having been used as a lookout by Indians waiting to ambush settlers coming down the Ohio River in flatboats. The tradition inspired a large wall painting in the lodge at Shawnee State Park located just a few miles away. The name comes from the imagined resemblance of the cliff to a large bird, Raven

Figure 115. Raven Rock Arch. The view of the arch as seen from above.

Rock itself being the head and beak, and the cliff walls on either side representing the outstretched wings.

Raven Rock Arch is found 30 feet southwest of Raven Rock (Figure 116). It takes the form of a thin remnant of stone 1.2 feet wide at its narrowest point between the edge of the cliff and an oblong skylight 15 feet long and 7 feet wide. This skylight formed in an unusual fashion (Figure 117). It appears to have begun as a small alcove in the face of the cliff which enlarged both inward and upward as a dome. The process most likely involved atmospheric erosion working on unequal cementation of the sandstone. As the dome enlarged upward, it broke through the surface of the upland just behind the edge of the cliff, leaving the inverted bowl's rim to form the arch.

The face of the cliff beneath the arch and Raven Rock is pitted, rounded, smoothed and etched into a number of interesting features, including several other domes similar to the one which formed the arch. In some places the sandstone contains very little cementing material and can be easily rubbed off with a finger. Given such weak rock, it is possible that wind has played a part in forming some of the small-scale surface features of the cliff. Cross-bedding was noted in several places and almost certainly played a role in determining their form.

Figure 116. Raven Rock Arch (foreground) and Raven Rock. Raven Rock extends out over open air beyond the arch. The Ohio River and the hills of Kentucky are seen in the distance.

Whatever processes formed the arch, they are still active. In a picture of Raven Rock Arch published at the turn of the twentieth century in *The Book of Ohio,* a similar arch is visible just beyond the main arch (Figure 118). This second arch has long since collapsed, although its curved back wall remains. On the opposite side of Raven Rock Arch, a large dome similar to the one responsible for the arch broke through to the upland sometime between the author's first and second trips, creating a funnel-shaped pit 1 foot across and 2 feet deep (Figure 119). Should this hole continue to enlarge, another arch similar in size to Raven Rock Arch could easily result. Thus, within a short stroll along the cliff top, the full life cycle of Raven-Rock-type arches can be seen.

Figure 117. Raven Rock Arch from below showing the upward-enlarging dome that broke through the surface of the ridge to create the skylight. The possible collapsed arch is seen just to the left of the present arch.

Against this dramatic evidence of ongoing arch evolution are several indications of stability. In the historic picture shown in Figure 118, the span and skylight of Raven Rock Arch appear almost exactly as they do today, more than a century later. Although the rock forming the cliff is weak in places, it is quite resistant in others where the cementing agent is stronger. Areas so protected appear to be especially durable. Many inscriptions carved into the rock over the last century-and-a-half appear to be as sharp as when first carved. One penciled message in a sheltered nook looks fresh enough to have been made yesterday, even though it is dated 1837! The great number of historic inscriptions testifies to the attractive power of this place. Two visitors making the trip were impressed enough to leave their inscribed thoughts in poetry:

Upon these majestic heights we stand,
And view the works of God and man.

213

Figure 118. Making unusual use of a natural arch — Raven Rock Arch in 1900 (Van Tassel, 1901). Note the apparent twin arch, at far left in the photograph, behind the dapper young gentleman.

The attraction of this location for the Indians is certainly understandable whether or not they actually used it as a lookout. It was more accessible to them than to us since two of their important trails, the Warriors Path from Lake Erie to Cumberland Gap and the Pickawillany Trail from the Shawnee villages at the mouth of the Scioto River to their town of Peckuwe near present Springfield, Ohio, met on the ridge behind Raven Rock. Unlike Euro-American settlers whose roads tended to follow river valleys, Indians preferred trails on the higher and drier ridge tops and divides. Tradition also tells us that Raven Rock was the site of magical contests between ancient medicine men whose endeavors so blasted the rock that only wiregrass will grow there. Supposedly one of the great leaders of these "Ancient Ones" was buried somewhere in the bowels of the hill in a tomb whose entrance is revealed now and then by a landslide, only to be conveniently covered up once again by another rock fall. That such

Figure 119. Raven Rock. A new skylight is forming a few feet from the existing one. The scale is 12 inches long.

catastrophic collapses do occur here is evidenced by a pile of large fallen rocks at the base of the cliff.

Far be it from me to slight accounts of strange forces at work in this place. After my first difficult ascent to the arch — "There is no easy way to the top of Raven Rock" advises a history of Scioto County published by the Portsmouth Recognition Society — I returned home to find that my camera had mysteriously quit working and all the pictures taken of the site had been lost. During a second trip to rectify the situation, a sudden storm with black clouds, howling wind and dramatic flashes of lightning sent me off the narrow, exposed ridge top at a run. Whatever power dwells there must have smiled on my third attempt, however, for the day was sunny, the camera worked fine and a pleasant hour was spent watching turkey vultures soaring at eye level. That day alone would be enough to convince any skeptic that there is magic in this place.

The presence of Raven Rock Arch and the other features associated with it is entirely dependent on the cliff in which they are found. Without it they would not exist, and the story of their formation is therefore tightly bound to that of the cliff. Before the advent of continental glaciation, this section of the Ohio River Valley was part of a broad upland between the Portsmouth River, which flowed through what is now the Scioto River Valley, and the Old Kentucky River which utilized the valley of the present Great Miami River. Both of these preglacial rivers flowed north to meet the main stem of the Teays River in west-central Ohio. Advance of the ice sheet blocked this northward drainage and the ponded flow, along with meltwater from the glacier, cut through the old drainage divides to create new outlets for the impounded water. The wide valley below Raven Rock Arch, one of these outlets, is now utilized by the Ohio River. The Logan Formation rarely forms cliffs except where rivers have undercut it and oversteepened the slopes beneath it. The cliff at Raven Rock Arch is apparently a relic of the end of the Pleistocene when the newly formed Ohio River, swollen with glacial meltwater, was much larger than the present stream.

In 1996, Raven Rock was dedicated as a state nature preserve. A small parking lot at the base of the hill and a very steep trail to the top of the ridge now provides public access. A free visitation permit from Ohio DNAP is required.

Rockgrin Arch (OH-A-SCI-02)* — Scioto County
Span: 5 feet (1.5 meters); Clearance: 1.25 feet (0.4 meter)

Rockgrin Arch (Figure 120) is found 26 feet below Raven Rock where the bottom of the cliff meets the top of the steep, forested slope which falls another 450 feet to the floodplain of the Ohio River. The arch has formed in a small buttress extending out from the base of the cliff. At 5 feet long and a shade over 1 foot high, this opening is near

Figure 120. Rockgrin Arch. This arch is located in the base of the cliff containing Raven Rock Arch.

the lower size limit of significant arches as defined for Ohio. The minimum size criteria for an arch to qualify for inclusion in the survey were, in fact, set so as to be able to include this arch.

The long, narrow opening which resembles a wide-mouthed grin takes up a good part of the buttress, leaving a lintel not much thicker than the height of the opening and an abutment only 9 inches wide on the downslope side. Atmospheric processes operating on rock with varying resistance carved this arch. Its floor and the area around it are covered with loose sand, the product of rock disintegration, and is a favored place for antlions to make their traps.

Slide Arch (OH-A-SCI-03)* — Scioto County
Length: 3.6 feet (1.1 meters); Clearance: 5 feet (1.5 meters)
Narrowing to a mere 2 inches thick, Slide Arch (Figure 121; Plate 12) possesses the thinnest lintel of any of Ohio's listed arches. Seen from above, it has a broad, flat appearance, being 10 inches wide through most of its length, although a notch cut into the front edge narrows it to 6 inches at that point. Like Rockgrin Arch, its close

217

Figure 121. Slide Arch. The view from the top of the "slide."

neighbor to the east, this arch is one of the state's smallest listed arches. It is also one of the most difficult to describe. The lintel continues the top edge of a 6-foot-high ledge of rock buttressing the bottom of the main cliff. An opening 3 feet wide and oval in outline has been carved through the ledge between its top and its face, leaving the lintel behind. This opening curves from horizontal to vertical in passing through the ledge, from perpendicular to the cliff face at the top of the ledge to parallel to it at its face. The result looks very much like a curving, round-sided slide which comes out through the arch entrance.

Atmospheric erosion working with differential cementation of the sandstone and guided by subtle bedding planes has played the largest part in forming this interesting feature. Wind may also have helped smooth the sides of the slide. Although no seeps or watercourses of any kind were noted near any of the Raven Rock arches, the notch in the front of the span of this arch may indicate that running water helped originate it.

The stability of such an arch is questionable. The fact that one arch on the cliff has apparently already collapsed does not inspire much hope that this thin span will last much longer. Even more sturdy features come to an end in time. Just beyond Rockgrin Arch to the east, a large slab of sandstone lies at the base of the cliff (Figure 221). In size, shape, color and hardness it is a twin to Raven Rock, jutting out against the sky above. This ancestral Raven Rock fell long before Euro-American settlers with their written records moved into the area. Raven Rock itself will follow in time, as will the less durable arches. There are few other places in Ohio that show the dynamic nature of geology so dramatically.

The Penthouse (OH-A-SCI-04) — Scioto County
Span: 3 feet (0.9 meter) Clearance: 1 foot (0.3 meter)
Located west of the other Raven Rock arches, The Penthouse (Figure 122) occupies a low outlier of the more massive cliff to the east. It is a small pillared alcove, and this rock-house form combined with its height above the Ohio River led to its name. It would be a

Figure 122. The Penthouse. The scale is 12 inches long.

small apartment, however, and a bit uncomfortable since the whole thing has a definite tilt toward the east. A slab-like pillar with an unusually ragged outline only 1.3 feet across at its widest point splits the front of the alcove into two entrances of decidedly unequal size. The smaller, uphill entrance is 2.5 feet wide and 1 foot high. The larger, downhill entrance is not much wider, but more than twice as high. Some interesting honeycomb texture formed by small-scale differential weathering was noted on the ceiling. There also was a disordered pile of sticks and grass cluttering a small ledge on the inside wall, remnants of a nest of some kind which shows that something found this dry, sandy apartment attractive.

Arches in Pennsylvanian Strata

In Pennsylvanian time, most of eastern Ohio was a geological battleground where shallow seas alternated with coastal swamps. When the sea won, marine shale, limestone, ironstone and flint were deposited. When the coastal swamps were in ascendance, nonmarine shale, sandstone, coal and limestone were laid down. Most of the plant fossils found in Ohio come from Pennsylvanian rocks which occupy a wide swath of the southeastern part of the state (Figure 7). The Sharon Formation contains most of the Pennsylvanian arches found in Ohio.

Sharon Formation

The Sharon Formation, the lowest, and therefore oldest, major Pennsylvanian rock found in the state, is one of the important arch-forming beds in Ohio. It surfaces in Scioto, Pike and Jackson counties in southern Ohio, and again in the northeastern part of the state where it forms the picturesque ledges which are such a notable feature of the area. Arches are found in both outcrops. The deposit varies between 10 and 250 feet in thickness, a result of its being deposited on the highly eroded, uneven surface of the Mississippian rocks beneath it. The Sharon is usually massive and cross-bedded, and varies from coarse-grained sandstone to true conglomerate. One of its most striking features is its high silica content evidenced by the lenses and layers of rounded quartz pebbles that in places make up the bulk of the rock. Most of them are one-half to three-quarters of an inch across, but occasionally pebbles as large as a hen's egg, and rarely some as large as a man's fist, can be found. Nearly all are white quartz, but rose quartz and even jasper pebbles are sometimes found. Such pebbles are typical of stream-worn gravel and this, combined with the cross-bedding, indicates that the Sharon represents a solidified delta (Stout, 1916).

Unlike the older Black Hand Sandstone delta which was built from highlands to the south, the Sharon Formation delta material was carried in from the east. The Appalachian Mountains were new then and a shallow sea washed against their western flanks. Streams carried eroded rubble from the mountains into the sea, depositing the delta that would become the Sharon Formation. Quartz survived the rough-and-tumble journey better than softer material and so formed the bulk of the growing delta. Iron compounds brought in by groundwater cemented the debris into a rock unit. Variations in the amount and strength of this iron cement play an important role in determining how the Sharon Formation erodes once it is exposed to weathering. On exposed surfaces, the oxidized iron forms a protective coating which gives the rock its deep-brown to reddish color.

One of the most interesting and accessible exposures of the Sharon Formation is found in Nelson-Kennedy Ledges State Park in Portage County. Here the characteristic jointing of the rock and its reaction to weathering are displayed in the large blocks which have separated from the main body of the ledge. In many places the underlying Mississippian shale has been eroded away, creating overhangs. The deepest of these, called Gold Hunters Cave, extends 80 feet into the bluff. The name comes from a short-lived gold rush in 1870 which ended when the "gold" proved to be iron pyrite — "fool's gold." Many of the separated conglomerate blocks have moved or are moving downhill, sliding ever so slowly on the surface of the shale beneath them. In some cases they have tipped away from the parent cliff, creating wide-topped gulfs. In others, they have leaned back against the cliff to form gravity arches, some of which are long enough to be locally called "caves." One of the largest of these is the Old Maids Kitchen (Figure 123) which is actually two gravity arches formed by leaning slump blocks located side-by-side. Where the blocks have not yet separated from the cliff, groundwater flowing through them has enlarged the crevices, a process which has been important in forming

Figure 123. Old Maids Kitchen in Nelson-Kennedy Ledges State Park. Shown here are two end-to-end gravity arches in the Sharon Formation.

arches in the northern outcrop. The wide stretch of flat ground between Nelson Ledges and Kennedy Ledges to the east is the valley of a preglacial river that was filled to a depth of 250 feet by glacial deposits, reminding us that Pleistocene events also played a part in the story.

Cats Den Natural Bridge (OH-A-GEA-01) — Geauga County
Span: 6.7 feet (2 meters); Clearance: 8.5 feet (2.6 meters)
Cats Den Natural Bridge (figures 124 and 125) is one of Ohio's more unusual natural bridges. It is the only one which actually carries a road, appropriately named Cats Den Road. Another unusual feature is that its skylight takes the form of a steep-sided pit, most likely the original "cats den" (Figure 125). This pit is 15 feet deep, 60 feet long

Figure 124. Cats Den Natural Bridge. A view through the opening from the pit.

and over 12 feet wide. It formed just behind a line of cliffs, one of northeast Ohio's typical "ledges," leaving a wall of conglomerate 21 feet wide. The opening of Cats Den Natural Bridge has been cut through this wall.

The formation of this natural bridge is intimately associated with the formation of its intriguing pit. The Sharon Conglomerate here is formed of a soft, easily eroded lower layer and a more-resistant cap layer. The lower layer is a conduit for groundwater, and springs and seeps are often found issuing from it. At Cats Den, a number of springs are found at the base of the wall of the pit opposite the arch opening. The vertical crevice which forms the opening of the natural bridge was no doubt the original outlet for all of them. As water passed through the rock toward this opening, it dissolved the cementing

Figure 125. Cats Den Natural Bridge. A view of the pit. The arch opening is barely visible on the right side of the photograph. The base of the road crossing the natural bridge can be seen above the opening.

material holding the sand and gravel of the conglomerate together. The extent of this weathering may have been increased by capillary action carrying moisture farther into the body of the rock.

A small, intermittent stream now falling into the north side of the pit also has been an important factor in forming it. The course of the stream across what is now the pit was controlled by a vertical crevice in the cap rock which runs nearly parallel to the face of the cliff and which can still be seen in the bed of the stream just above the point at which it falls into the pit. Water from the stream seeping down through this fracture entered the porous lower layer, increasing the level of moisture within it and adding to the flow of groundwater utilizing the opening in the face of the cliff. In time, much of the flow of this stream was captured by this combination of fractures. The result of all this groundwater flow and accompanying solution of cement was that the lower layer of conglomerate was reduced to its constituent sand and pebbles along the route of flow. This rubble was then flushed out through the enlarging crevice in the face of the cliff.

As the lower rock disintegrated, the cap rock lost its support. Eventually it separated along vertical crevices cutting across it and collapsed into the embryonic pit. This opened a way for the stream to fall directly into the pit. Its then-unobstructed flow rapidly flushed out much of the accumulated disintegration debris of the lower layer while at the same time breaking down the fallen roof blocks and enlarging the opening of the incipient natural bridge.

With the stream now falling into the pit, more water was available for attacking its walls. Moisture from the falls would have been held within the pit's narrow confines to condense on the surrounding rock, furthering its disintegration. The stream itself, flowing along the floor of the pit, would have attacked the vulnerable lower layer, especially in times of flood. The springs also continued their work, although now they came from the back wall of the pit where they formed deepening alcoves before joining the stream to flow out through the face of the cliff. As more blocks of undermined cap rock gave way, the pit enlarged to its present size.

This history is quite visible in the present configuration of the pit. Its outline along the top edge is very angular and obviously joint-controlled, a result of blocks of cap rock separating along joint planes and falling into the pit. The outline of the floor of the pit is, by contrast, more rounded and also much larger. The walls of the pit in the bottom layer have been cut back extensively beneath the cap rock on every side except the south. Springs flow from deep alcoves, and small arches have formed in some of the fins of rock left between them.

The difference in resistance to erosion of the two rock layers is well illustrated in the bridge opening which was formed by the widening of another vertical crevice, this one perpendicular to the cliff face (Figure 124). In the lower, softer layer of conglomerate, the opening has a wide, ragged outline. In the hard upper layer, it is straight-sided and very narrow. Construction of the road across the top of the arch has obscured and perhaps changed the surface of its lintel, making it difficult to determine whether or not this crevice was originally open to the sky. Regardless of the truth of the matter, Cats Den Natural Bridge is such an interesting geological feature that it behooves us to give it the benefit of the doubt.

Present erosion rates appear to be low. There may have been a time, perhaps during the retreat of the Wisconsinan glacier, when the stream carried more water and had more cutting power. Since this part of Ohio was completely covered by Wisconsinan ice, the natural bridge could not have started forming until its retreat. Cats Den Road now leads through an area of extensive housing. Much of the ground has been graded in the recent past and so it is difficult to tell what the original drainage pattern of the stream leading into the pit might have been. Given its configuration, however, it is probable that drainage lines have not been greatly changed.

Although no confirmation has yet been found, it is assumed that the name "Cats Den" refers to mountain lions (cougars or panthers, *Felis concolor*) which would have been an important predator in the area during the time of early Euro-American settlement. The pit is probably the "cat's den," so called either because mountain lions were actually found there or, more likely, because it looked to pioneers like a place where they should have been found. It is unlikely that any self-respecting cat would long occupy such a damp hole.

Camp Christopher Natural Bridge (OH-A-SUM-01)*
— Summit County
Span: 5 feet (1.5 meters); Clearance: 20 feet (6.1 meters)

Camp Christopher Natural Bridge (Figure 126) takes its name from the Catholic Youth Organization's Camp Christopher (named for Christopher Columbus) where it is found. It resembles Cats Den Natural Bridge in that it formed in a wall left between a pit in the Sharon Formation and a line of ledges. This pit is deeper than that at Cats Den, however, dropping 23 feet. Its other dimensions, 60 feet long and 20 feet wide, are comparable to the Cats Den pit. The conglomerate here appears to be more resistant to erosion through its entire exposed depth. The walls of this pit are straight-sided with very little undercutting and the enlarged vertical crevice which forms the opening of the natural bridge exhibits no marked widening in any part.

Figure 126. Camp Christopher Natural Bridge. A view through the opening from the pit.

This bridge was formed as a result of erosion by an intermittent stream falling into the pit. The stream, the opening of the natural bridge and the valley beyond it are all in line. A groove across the top of the lintel most likely represents the original course of the stream to the edge of the cliff. Eventually it was detoured through crevices down into the enlarging pit and out through the vertical crevice in the face of the cliff which was then enlarged to form the arch opening. This widening extends to within 1 foot of the top of the lintel. The means by which the pit formed are not clear. Undercutting and collapse may have played a part, but that is not as apparent here as at Cats Den Natural Bridge. The uncertainty is due in part to a pile of glacial erratics dumped into the pit down which the stream now drops as a cascade. The boulders were no doubt brought to the pit and dumped into it by farmers clearing the surrounding fields. Perhaps they felt that the pit was useless and dangerous, and therefore a good place to dispose of their boulders.

The surroundings of Camp Christopher Natural Bridge retain their primeval, forested aspect, unlike those of Cats Den Natural Bridge which is hemmed in by houses and the bleak, clear-cut swath of a major electrical transmission line. The short, steep-sided gorge leading from the arch opening is especially impressive, shaded as it is with Eastern Hemlock and other northern-type trees.

Bass Lake Natural Bridge (OH-A-GEA-02) — Geauga County
Span: 13 feet (4 meters); Clearance: 4 feet (1.2 meters)

Although formed in the same northern outcrop of the Sharon Formation as Cats Den and Camp Christopher natural bridges, Bass Lake Natural Bridge (figures 127 and 213) does not share the unusual configuration of the other two. Instead, the blocky, bridge-like

Figure 127. Bass Lake Natural Bridge. A view of the lintel from the pit behind the arch. The importance of joints in forming this feature is well shown by the flat back of the lintel and the right angle formed where it meets the wall of the pit.

PLATE 1

Plate 1. The massive lintel of Miller Arch is emphasized in this view from the floor of Rocky Fork Gorge.

PLATE 2

Plate 2. Castlegate Arch was named for its resemblance to a passage through an ancient castle wall.

Plate 3. Trimmer Arch exhibits the most perfectly arch-shaped opening of any natural arch in Ohio.

Plate 4. The tenuous connection between Greenville Falls Arch and its parent cliff is apparent when viewed from inside the skylight.

Plate 5. The eastern entrance of Ohioview Arch frames a view of the Ohio River Valley and the Kentucky hills.

Plate 6. The full depth of the alcove that was breached to form Rockbridge is visible in this view.

Plate 7. The waterfall responsible for forming Balcony Natural Bridge usually flows strongest in the spring. In this view, the lower falls is cascading off the "balcony."

Plate 8. The upstream side of Hagley Hollow Arch exhibits the subtle bedding often found in the massive layers of the Black Hand Sandstone.

Plate 9. The remnant block of Black Hand Sandstone through which most of Rock House has been eroded is readily visible in this view of its east entrance.

Plate 10. A view of the colorful exterior of Rock House showing a portion of the rock wall separating two of the feature's windows.

Plate 11. Raven Rock Arch has a massive, solid appearance, but the joint cutting across the lintel in the shadowed area will eventually prove to be a fatal weakness.

Plate 12. Like Raven Rock, Slide Arch overlooks the Ohio River Valley and the distant hills of northern Kentucky.

Plate 13. Like most of Ohio's natural bridges, Lucas Run Natural Bridge crosses what is a dry valley for much of the year.

Plate 14. Seen from below, Ladd Natural Bridge is a massive, brooding presence.

Plate 15. The small skylight of Cobble Arch is barely visible on the left side of the dark main opening. Natural arches formed in cemented glacial gravel are rare.

Plate 16. John Bryans Window as seen from below. The "window" separating this tea table from the cliff opens behind its narrow leg.

appearance of Bass Lake Natural Bridge is common to most other natural bridges in the state. Found at the end of a low-walled box canyon, it appears to be the result of a breached alcove formed from waterfall sapping. Viewed from the front, it presents the typical thick lintel with a gently curved underside found in most natural bridges formed in this manner. The back wall of the lintel is straight, showing evidence of joint control. The front wall was less affected by joints and is more ragged in outline. The lintel is 3 to 4 feet thick and varies in width from 5 feet at the southwest end to 3 feet at the northeast.

The bottom of the lintel is made of a layer of conglomerate that is more resistant to weathering than those layers above and below it. This resistant layer appears to hold up the bridge and is probably the major structural element involved in its existence. Although this layer is 1.4 feet thick in the northwest abutment, it thins to only 8 inches at the center of the bridge. This difference is a result of thinning from below due to the loss of material by sapping during formation of the arched ceiling of the original alcove.

The skylight, nearly 13-feet square, is quite angular where its rock sides are visible and evidently owes much of its form to joint control. It appears to be another example of a pit cut into the Sharon Formation behind a natural bridge, much like those at Cats Den and Camp Christopher natural bridges, only more shallow.

Apparently this arch began as an alcove cut back into the low ledge now marking its front by the small stream which formed a waterfall at the site. In a repetition of the story found at so many of Ohio's natural bridges, water from the stream seeping down through crevices in the bedrock and out through the face of the waterfall not only enlarged the alcove, but also widened the crevices until the entire flow of the stream dropped behind the face of the ledge and out beneath the more-resistant top layer which subsequently became the lintel of the natural bridge. Since then, the skylight has been enlarged to its present size.

The total height of this bridge from the stream bed to the top of the lintel is only 7 feet. Even so, its classic shape and the light it sheds on larger natural bridges of the same type in Ohio give it a value beyond its size. Bass Lake Natural Bridge was accidentally discovered by Michael Hansen of Ohio DGS while mapping outcrops of the Sharon Formation in the 1980s. Although local residents knew of its existence, they had not realized its importance. In the years since it was discovered, the arch has been encroached upon by an expanding quarry (Figure 213). Although the natural bridge itself is still intact, half of the pit behind it now lies beneath a high berm of rock and earth which hides the quarrying operations beyond. The small stream which once fed into the pit and was responsible for forming the arch has been obliterated.

Pike Arch (OH-A-PIK-02) — Pike County
Span: 30 feet (9.1 meters); Clearance: 10 feet (3 meters)
Massive Pike Arch (Figure 128), found in the southern exposure of the Sharon Formation, can be considered Ohio's major example of arch formation through headward erosion. In this case, two tiny, intermittent streams have pierced a rock buttress 60 feet wide and extending 80 feet out from the side of a small, steep-sided valley. This buttress is located near the rim of the valley and the two streams head very near it on either side.

Figure 128. Pike Arch. This is the largest example of a headward erosion type of breeched-alcove arch in Ohio.

As these streams cut their own short, shallow valleys, they encouraged the formation of alcoves on either side of the buttress. Eventually, the deepening alcoves broke through their common back wall and formed the arch. This is quite evident from the floor plan of the arch which is narrowest at the center where the wall separating the two shelters originally stood. The shape of the opening has been determined by both unequal cementation of the conglomerate and sediment grain represented by cross-bedding. Collapse has also played a part as shown by the breakdown rubble littering its floor. The largest of these rock slabs is 33 feet long, 19 feet wide and 9 feet thick — not something to be near when it fell. Several of these slabs exhibit graffiti, most of it of the recent coarse type too commonly found in such places. However, a few of the older petroglyphs took time and careful work to complete and approach the status of folk art. Among these are depictions of a man's face, a squirrel and several arrowheads. The oldest dated work is from 1909.

This is one of Ohio's larger arches. Its span, measured at the narrowest point where the common wall between the original alcoves was breached, is 30 feet. The opening is 63 feet long from entrance to entrance and has a clearance at its lowest point of 10 feet. It rapidly increases to twice this height at each entrance. The buttress itself is 25 feet high at the arch center, giving the lintel a thickness of 15 feet. This massive lintel combined with the large abutment on the downslope side gives Pike Arch much of its impressiveness.

The top of the buttress and adjoining slopes support a sparse pine forest, its floor speckled with Reindeer Moss, actually a lichen (*Cladonia* spp.), which is typical groundcover on such thin soils. Sand mining in the valley of Big Run below the arch came within several hundred yards of it. Long abandoned, the former quarry provides an unusual sandy environment which supports a developing plant community of some interest. As would be expected in such a place, Ohio's only native cactus, the Prickly Pear (*Opuntia humifusa*) grows here, as does the more gentle Narrow-leaved Blue-curls (*Trichostema dichotomum* var. *lineare*) and the Round-leaved Catchfly (*Silene rotundifolia*) which grows on the arch itself. All these plants are rare in Ohio, but even more rare is Buffalo

231

Clover (*Trifolium reflexum*) which grows here and nowhere else in the state. This interesting plant community and perhaps the arch itself are unfortunately still endangered by the recurring possibility of renewed mining.

The Sharon Formation makes its presence known in several impressive geological features found within a few miles of the arch — Whites Gulch, Chimney Rock (Figure 179) and, most notably, Big Rock (Figure 129), a massive, sheer-sided outcrop rising like a miniature mountain above Dry Run Road.

The name Pike Arch was given in reference to the feature's location in Pike County and is somewhat unfortunate in that two other arches have since been reported from the same county. The name Pike Arch has been retained, however, because of its frequent use in previous publications.

Figure 129. Big Rock offers an impressive exhibition of the southern exposure of the Sharon Formation.

Hole-in-the-Wall Arch (OH-A-JAC-01) — Jackson County
Span: 3.9 feet (1.2 meters); Clearance: 1.8 feet (0.5 meter)

Located in a neighboring valley just east of Pike Arch, Hole-in-the-Wall Arch (Figure 130) represents an earlier stage in the same process that formed the larger arch. Here two alcoves have recently, in geological terms, joined. The hole in their common wall is still small, being only 3.9 feet wide and 1.8 feet high. There is a vast difference in size between the two alcoves which have joined. The larger one has a reach exceeding 25 feet and a depth of 22 feet. In contrast, the smaller alcove has a reach of only 11 feet and a depth of 8 feet. The opening between them breaks through the back wall of the smaller alcove and a side wall of the larger. It is easy to see how continued thinning and collapse of the common wall between the alcoves over time will enlarge the opening.

This particular exposure of the Sharon Formation contains lens-shaped masses of coarse pebbles embedded in a matrix of sandstone. These lenses exhibit great variation in thickness laterally, narrowing from 10 feet or more to the thickness of a single layer of pebbles within the distance of a few steps. Most of the pebbles are of quartz and range in size from a quarter of an inch to 3 inches in length. All

Figure 130. Hole-in-the-Wall Arch. The entrance of the large alcove is on the left.

exhibit the well rounded outlines of water-worn gravel common to the Sharon Formation. These pebble lenses appear to have influenced the location of niches and alcoves carved into the sandstone bluffs of the region. The widest part of Hole-in-the-Wall's opening occurs in a thin layer of pebbles. A large alcove above and just to the north of the small alcove of the arch has formed in a more extensive layer of pebbles. Although the pebbles appear to be strongly cemented into the matrix of the rock, the coarse texture of the lenses they form provides a more efficient conduit for the passage of groundwater than does the finer-textured sandstone surrounding them. This would lead to a more active solution of the cementing material holding the lenses together and result in their more rapid disintegration. The fall of a loosened pebble the size of a small egg will naturally create a larger void than would the fall of a loosened grain of sand. Thus the pebble lenses weather out more rapidly, resulting in the alcoves and niches which form such a notable feature of the Sharon Formation outcrops in this part of the state.

At one time the conglomerate in this region was mined for its sand and gravel. The large expanses of gravelly sand found in the larger valley below the small tributaries containing both Pike Arch and Hole-in-the-Wall Arch remain as evidence of this past use. Fortunately, that effort has not yet extended to the arches themselves, although continued interest in reopening the mine makes that sad result a definite possibility.

Lake Katharine Arch (OH-A-JAC-02) — Jackson County
Span: 15 feet (4.6 meters); Clearance: 2 feet (0.6 meter)

Lake Katharine Arch (Figure 131), located deep within the wilds of Lake Katharine State Nature Preserve near Jackson, is a typical example of the enlarged-joint-and-bedding-planes process of arch formation. The opened joint is narrow and very angular, being only 4 feet wide at the top and narrowing toward the bottom. The lintel itself has a width of 9 feet and varies in thickness from 3 feet at the rear to 2 feet at its front edge. The alcove eroded from the bedding plane has a reach of 27 feet.

The arch is located in a line of sandstone outcrops near the top of a steep valley wall dropping into one of the tributaries of Rock Run. The

Figure 131. Lake Katharine Arch. The massive lintel contrasts noticeably with the small skylight.

hillside above it has a shallow gouge funneling into the skylight which shows that some surface water passes beneath the arch. This wash, slight though it undoubtedly is, has most likely been the most effective factor in enlarging the crevice to form the skylight. Since the back wall of the crevice is also the back wall of the alcove, it appears that crevice enlargement and alcove deepening took place at the same time.

Although the arch is not publicly accessible, the main part of the preserve on the opposite side of Lake Katharine is open for visitation and has an extensive network of trails with interpretive signs. By building a dam across Rock Run, Edwin A. Jones and James J. McKitterick, the original owners, created an impressive clear-water lake surrounded by sandstone cliffs and wooded hills that became the centerpiece of a summer camp they operated for many years. In a generous display of foresight, they eventually donated the area to Ohio DNAP. The rich Appalachian forest of the preserve protects extensive stands of Eastern Hemlock (*Tsuga canadensis*) and Ohio's largest population of Umbrella Magnolia (*Magnolia tripetala*). Here, too, is the state's only stand of Bigleaf Magnolia (*Magnolia macrophylla*), growing at its northern limit and listed as endangered in Ohio.

Peephole Arch (OH-A-JAC-03) — Jackson County
Span: 3 feet (0.9 meter); Clearance: 1.8 feet (0.5 meter)

Peephole Arch (Figure 132) was named for a nearly circular hole piercing its outer abutment. This abutment takes the form of a ragged column connecting the floor and ceiling of a shallow alcove found in the same low rock face containing Lake Katharine Arch. The alcove and the opening of the arch were created by atmospheric weathering; the column, being more strongly cemented, has greater resistance to weathering and so has endured.

Figure 132. Peephole Arch. The "peephole" is the circular hole visible in the ragged column defining the right side of the arch opening. The oval opening beyond Peephole Arch is too small to qualify for listing as an arch. Lake Katharine Arch is out of view to the right.

Arches in Other Pennsylvanian Strata

Woodbury Natural Bridge (OH-A-COS-01)*
— Coshocton County
Span: 5 feet (1.5 meters); Clearance: 2.2 feet (0.7 meter)

Woodbury Natural Bridge (figures 133 and 134) is a surprise. It is found in a part of the state where coal-rich Pennsylvanian sandstones form rolling hills. A natural bridge located here would be expected to have formed in sandstone, but Woodbury Natural Bridge is not a sandstone feature. It is formed of limestone and occupies center stage in a 50-foot-square piece of mini-karst.

The explanation for this apparent anomaly lies in the Allegheny Group of Pennsylvanian rocks which form the bedrock of the area. Composed mainly of alternating layers of sandstone, clay and coal, the Allegheny Group also contains a few layers of limestone. One of these, the Putnam Hill Limestone, outcrops in the bed of an unnamed tributary of Simmons Run in Woodbury Wildlife Area. The rocks in this small valley dip gently downward in an upstream direction, causing the tributary to cut across successively older strata of the Allegheny

Figure 133. Woodbury Natural Bridge. A view of the front of the arch from downstream.

Figure 134. Woodbury Natural Bridge. Vertical fluting cut into the back wall of the skylight by the waterfall — a common feature of karst terrain — is shown in this photograph. The scale is 12 inches long.

series. When it reached the Putnam Hill Limestone, the stream found a rock more resistant to erosion than the others in its path and formed a waterfall 6 feet high. In this place the Putnam Hill Limestone is made of a series of thin layers, each only 1 or 2 inches thick except for the top one which forms a resistant cap layer 1 foot thick. This cap rock is responsible for the greater resistance to erosion of the entire bed, as is shown by the shallow recess eroded into the weaker layers of limestone beneath it at the falls.

Nine feet behind the original position of the waterfall, a vertical crevice cuts through the limestone perpendicular to the course of the stream. Water seeping into this crevice and then out to the face of the falls between the thinner rock layers below the cap rock dissolved the soluble limestone. As the crevice and bedding plane openings were enlarged, more of the stream's flow was diverted through them which, in turn, widened the opening even faster. Eventually the entire stream was captured by the crevice, now enlarged into a skylight 10 feet long and 4 feet wide. The falling water has since cut a notch into the back wall of the skylight, carving in the process a small but beautiful display of fluting into the low limestone wall (Figure 134). The crevice itself has been enlarged to the south, passing under cover of the hillside as a narrow cave 2 feet wide, 2 feet high and 20 feet long.

In seeping out through the bedding planes of the thin limestone layers, the stream widened them and flushed out the intervening rock. Because these bedding planes have a slight downward tilt in the upstream direction, gravity would not have been much help in moving the water through them. Hydrostatic pressure created by the weight of water filling the enlarging crevice, aided in the early stages by capillary action, could have forced the water up the slight grade of the bedding planes and out the back wall of the recess behind the old waterfall, dissolving and weakening the thin limestone layers. This process is still occurring, for water, no doubt captured by crevices farther upstream, can be seen seeping out from between these same bedding planes in the upstream wall of the skylight.

Beaver Creek Natural Bridge (OH-A-COL-1)
— Columbiana County
Span: 8 feet (2.4 meters); Clearance: 3.7 feet (1.1 meters)

Although small, Beaver Creek Natural Bridge (figures 135 and 136) is instructive. In fact, its diminutive size enables an observer to comprehend the process of its formation more easily than can be the case with larger, more overpowering features.

Beaver Creek Natural Bridge was discovered by James Murphy, one of the earliest compilers of natural bridge information in Ohio, around 1970. He named it for the North Fork of Little Beaver Creek into which the unnamed stream flowing under the bridge falls a short distance to the west. The valley of this small stream has been cut into the series of shales, clays, limestones and sandstones of the Allegheny Group. At the level of the bridge, it is passing over a thin layer of resistant Lower Freeport Sandstone.

As is usually the case in such instances, the resistant sandstone forms the lip of a small waterfall which has been maintained by sapping of the underlying weaker rock and collapse of part of the unsupported lip. Eventually, the water of the stream found an outlet through a crevice in the sandstone upstream from the falls. In making its way

Figure 135. Beaver Creek Natural Bridge. The arch as seen from downstream. The scale is 12 inches long.

down through the crevice and out the face of the falls through the thin-bedded rock below, it enlarged the crevice enough to take all the water of the stream. The thin layers of rock beneath the lip of the falls were carried away, leaving the resistant layer of sandstone extending across the stream as a natural bridge.

The lintel of this bridge is a tabular sheet of sandstone 4 feet wide and only 9 inches thick, giving it a delicate appearance. The sides of the original alcove beneath the bridge are cut back deeply, giving it a reach of nearly 20 feet. Fifty yards upstream, the thin beds of rock visible at the side of the stream arch up to form a small anticline 10 feet across (Figure 136).

Beaver Creek Natural Bridge stands at the very edge of glaciation in Ohio. Thin remnants of Titusville Till of Early Wisconsinan age are found on the surrounding hills. The valley of the North Fork of Little Beaver Creek in the vicinity of the natural bridge is narrow and gorge-like with steep walls rising 300 to 400 feet and standing in places less than a quarter of a mile apart. This shows evidence of the glacial diversion of pre-existing drainage over a divide. The small valley containing the natural bridge has been eroded into the east side

Figure 136. Beaver Creek Natural Bridge. The small anticline in the valley wall above the arch is clearly evident in this photograph. The scale is 12 inches long.

of this gorge and so the natural bridge could not have formed before the gorge was cut sometime during the Pleistocene.

In 1764, Colonel Henry Bouquet marched a contingent of His Britannic Majesty's troops into what was then Indian country to over-awe the warring Shawnees on the Muskingum and Scioto rivers. The Great Road which the army followed crossed the North Fork of Little Beaver Creek near the mouth of the unnamed stream flowing under the natural bridge. It is doubtful that any of the soldiers noticed this little whimsy of Nature since wandering about the countryside look-ing for things of interest was almost certainly not encouraged. No record of its having been seen at such an early date has been found; indeed, the first record of its having been discovered at all is Murphy's report.

241

Mineral Arch (OH-A-ATH-01) — Athens County
Span: 25 feet (7.6 meters); Clearance: 5.5 feet (1.7 meters)

Mineral Arch (Figure 137) is located near the base of a hill where a resistant layer of sandstone crops out. It is a breeched-alcove arch which formed when a vertical crevice in the roof of the alcove was widened by erosion. The rock forming the arch appears to be very strong. The rounded arch shape of the opening is an expression of a tension dome and is a common form in arches found in sandstones.

This arch received its name from the town of Mineral located a short distance to the west. Originally called Mineral City, it was laid out by the Marietta and Cincinnati Railroad Company in the late 1800s. The mineral referred to was coal. According to a history of Hocking County written shortly thereafter, "The future of the town does not come under the head of Great Expectations, but as it is a great convenience the citizens and farmers around are satisfied" (Interstate Publishing Company, 1883). This less-than-thundering affirmation has been amply fulfilled. The town remains a small crossroads, but it does

Figure 137. Mineral Arch. This view shows the arch as seen from below.

remain, and it does have at least two geological claims to, if not fame, at least notice — Mineral Arch east of town and, on a hill overlooking it, Mineral Tea Table, described below in the catalog of Ohio's natural pillars.

Arch Rock (OH-A-VIN-01)* — Vinton County
Span: 9.7 feet (3 meters); Clearance: 8.25 feet (2.5 meters)
Arch Rock (Figure 138) is appropriately named, for this feature definitely resembles an arch. Although not large even by Ohio standards, its picturesque wooded location high above Elk Fork in the hinterlands of Vinton County makes it especially attractive. It has formed in sandstone of the Conemaugh Group which caps the ridges of the region. The opening of the arch pierces a fin of rock located at the end of one of these narrow ridges and appears to have formed through atmospheric weathering. This assumption is strengthened by the presence of several openings piercing the downstream abutment. The largest has a diameter of 1 foot, forming a small "window." The

Figure 138. Arch Rock. Although not large, this arch is notable for its delicate form.

243

abutment itself is about 4 feet thick and has the pitted, rugged surface expected of rock that has been eroded by the weather for a long period of time. The upstream abutment of the arch is part of the main body of the cap rock topping the ridge.

The lintel is approximately 4 feet thick. Although this is large in relation to the clearance of the opening, Arch Rock still manages to project a surprisingly graceful appearance. The sandstone is spotted with dark brown iron concentrations, the most notable being a 6-inch-thick band near the bottom of both abutments. Due to its greater resistance to erosion, this band of rock has not weathered back as much as the sandstone enclosing it and so forms a very narrow ledge. Although such iron concentrations are now just a geological curiosity, in an earlier day they held great industrial promise. From 1818 to 1916, the Hanging Rock region between Logan, Ohio, and Mount Savage, Kentucky, supported 46 blast furnaces that produced iron from local ore deposits such as this. Fired by charcoal made from trees stripped from the surrounding hills, these furnaces played a major role in the industrial growth of the United States. They were especially important during the Civil War when they supplied iron for the Union war effort, including the metal plates sheathing the *Monitor*. With the discovery of richer iron deposits in the Lake Superior region, the Hanging Rock furnaces were abandoned. The ruins of nearby Vinton Furnace can still be seen. Both the furnace site and Arch Rock are located within the Vinton Furnace Experimental Forest and can be visited by making prior arrangements.

Liberty Natural Bridge (OH-A-WAS-01) — Washington County
Span: 6 feet (1.8 meters); Clearance: 5 feet (1.5 meters)

The picturesque Liberty Natural Bridge (figures 139 and 140) is hidden on an unnamed intermittent tributary of an also unnamed tributary of Goss Fork. It was found by Marilyn Ortt in 1976 and named Liberty for both the township in which it is located and the bicentennial year of the United States in which it was reported. It has formed in a sandstone unit of the Monongahela Group, probably the Sewickley Sandstone, and resulted from the breaching of an alcove carved out by the intermittent stream beneath a layer of resistant rock. This alcove presently has a reach of 32 feet and a depth of approximately 14 feet.

Figure 139. Liberty Natural Bridge. A view of the arch from downstream.

The skylight cut by the stream forms a near perfect isosceles triangle with base and altitude both equaling 6 feet. It is somewhat unusual in that the apex points downstream and the waterfall occupies the wide base. Normally, the actively eroding side of a skylight is narrower than the side which forms the rear of the natural bridge. This backwards configuration gives the lintel a bowtie shape when viewed from above — narrow at the center and wide on each end. At its narrow center, the lintel is 5 feet across and only 1 foot thick, giving this natural bridge a delicate appearance which is augmented by its coy hiding place behind a massive slump block.

Although the narrow valleys of the unnamed tributaries of Goss Fork are now reverting to woodland, they once were occupied. Downstream from the natural bridge lie the remnants of a barn and house. Their carefully cut foundation stones and a sturdy fireplace are all that remain. A massive stone retaining wall along the creek which runs between them is reported to have been a response to the devastating 1913 flood which damaged the barn, forcing its owner to shelter his cattle in one of the large alcoves found nearby.

This valley is quite narrow and flat land for crops and pasture is very limited. That such impressively sturdy structures should have

Figure 140. Liberty Natural Bridge. Convenience trumps esthetics: an old oil pipeline has been placed on this natural bridge to cross the stream.

been built here indicates that a subsidiary source of income might have been available. Rusty, pitted iron pipes running along and across the bed of every stream in the vicinity may provide an answer. Shortly after Colonel Drake drilled his pioneer oil well near Titusville, Pennsylvania, oil fever struck eastern Ohio. Noting the presence of oil seeps in southern Morgan County, would-be oil barons opened the Macksburg Field with enough success to encourage further explorations in the region. Liberty Township in Washington County soon became part of the boom. Farms were leased by drilling companies for what then was considered an enormous amount of money. In those pre-Environmental Protection Agency days, wells were drilled wherever a derrick could be erected and pipes to convey the resulting oil were laid with an eye strictly to convenience and profit. One was even laid over the natural bridge (Figure 140). Since early oil strikes were made in the bottom of stream valleys, these were considered the most likely places to drill. It did not take long for the limited oil of the region to play out. The derricks and wooden structures associated with drilling have for the most part disappeared. The iron pipes remain, however, along with an abandoned boiler lying in the stream bed as a reminder of a time when the dream of making a fortune overnight came to the hills of Washington County.

Lucas Run Natural Bridge (OH-A-MRG-01) — Morgan County
Span: 27 feet (8.2 meters); Clearance: 11 feet (3.3 meters)

Although a medium-sized natural bridge for Ohio, Lucas Run Natural Bridge (Figure 141; Plate 13) is impressive for its solid massiveness and definite bridge-like character. The lintel is only 7 feet wide at its narrowest point, but it is also 7 feet thick, and its deck is flat enough to give a pedestrian confidence. It has been carved from a massive sandstone layer in the upper part of the Monongahela Group.

Figure 141. Lucas Run Natural Bridge. This view shows the upstream side of the arch. Marilyn Ortt, Ohio DNAP, provides scale.

Located near the head of a side stream falling into an unnamed tributary of Lucas Run, this natural bridge takes its name from the larger stream. It resulted from the breaching of an alcove which presently has a reach of 60 feet. The breaching took place along a steeply sloping joint which has determined the location of the back face of the lintel. The angle of this face corresponds to the slope of the joint which is visible in the walls of the alcove on either side of the span. Shallow, weathered honeycombing on this face may have formed initially during the seepage phase as water from the stream worked its way down into the crevice.

Since the intermittent stream breached the alcove roof on the line of the joint it has cut its way back into the hillside, creating an elongated skylight with steeply sloping sides. At the rear of the skylight, the stream cascades down a series of low ledges rather than falling off the cut-back edge of resistant rock forming the roof of the alcove — as happens in most breached-alcove arches. This sloping back wall of the skylight may indicate that all the layers of rock have more resistance here than they did at the face of the falls. It could also be a result of erosion over a greater length of time than has been available at most other natural bridges of similar form in Ohio. If so, then Lucas Run Natural Bridge may be one of the older such features in the state. Detailed study of the drainage history of the immediate area could provide some answers.

Although somewhat removed from the more traveled routes of the region, Lucas Run Natural Bridge has had its share of visitors, as shown by the names carved into it. The only dated entry noticed was from 1881. The natural bridge was described in Howe's *Historical Collections of Ohio*, published at about that time, as:

> . . . a huge stone arch, spanning a hollow which forms a rocky channel, sometimes dry and sometimes swollen by rains. Over the arch a grapevine runs riot, and here and there dainty fringes of cool ferns cling to the damp earth near its extremities. Underneath, the walls are covered with the initials of stragglers, who seek enduring fame after the manner of visitors to such spots. The bridge is perhaps thirty feet from end to end, fifteen feet high, and so wide as to allow a sleigh to cross with safe margin (Howe, 1896).

It is interesting to note the close agreement of measurements made more than a century apart. The difference between Howe's "thirty feet from end to end" and the more recent 27-foot measurement of its span can be readily accounted for by assuming that he or his informant were not so concerned about measuring the most constricted part of the span. His 11-foot height matches the recent measurement of clearance exactly. At 7 feet wide, the lintel could certainly support a sleigh so long as the driver was careful.

A small alcove in an adjoining hollow to the east has been given the name Horse Thief Cave. The operation must have been a minor one, however, for both the cave and the hollow are large enough to house only a few horses and thieves.

Arches in Permian Strata

Ohio's Permian strata are the youngest bedrock found in the state. They occupy a narrow band along the Ohio River between Meigs and Jefferson counties (Figure 7). The sea level fluctuations that resulted in the repeating cycles of sandstone, shale, freshwater limestone and coal deposition during Pennsylvanian times continued into the Permian. Extensive swamps harbored amphibians and reptiles, represented in the rocks by fossilized remains and occasional preserved footprints. Two of the most interesting of these animals are the sail-backed pelycosauran reptiles *Dimetrodon* and *Edaphosaurus*, familiar to all fans of prehistoric reptiles. By mid-Permian times, the surface of Ohio was above sea level and has remained there ever since. During the long interval of Mesozoic and Tertiary times, the age of the dinosaurs and mega-mammals, Ohio was undergoing erosion rather than deposition and no rock strata, and therefore no fossils, were laid down.

Although southeastern Ohio was not covered by the Pleistocene glaciers, it was affected by them. The creation of the Ohio River by glacial meltwater cutting through drainage divides between preglacial rivers effectively lowered the base-level of the area, and pre-existing streams as well as meltwater rivers streaming off the retreating glacier cut their beds down to meet it. These deepening valleys created cliffs and ledges which provided potential sites for natural arches.

249

Mustapha Natural Bridge (OH-A-ATH-02) — Athens County
Span: 35 feet (10.7 meters); Clearance: 11 feet (3.3 meters)

Mustapha Natural Bridge (figures 142 and 143), named for an island in the Ohio River located a short distance downstream from it, has formed in the Hockingport Sandstone Lentil, a unit of the Dunkard Group. This sandstone was laid down by north-flowing streams meandering across a gently sloping alluvial plain. It is a small deposit only 30 miles long and 10 miles wide. It does, however, reach a thickness of nearly 100 feet in the vicinity of Mustapha Natural Bridge.

Figure 142. Mustapha Natural Bridge. This view taken underneath the arch shows its massive form. The Ohio River, not visible in this photograph, is to the right.

Figure 143. Mustapha Natural Bridge. A view of the Ohio River from within the alcove.

This arch has formed where a small, intermittent stream comes down the face of a bluff on the north side of the valley of the Ohio River, falling over a ledge of resistant rock and eroding a deep alcove into the weaker layers beneath it. A crevice in the roof of the shelter has been subsequently enlarged by seepage from the stream passing over it and now carries its entire flow. The rim of the alcove remains as the lintel of the natural bridge and is 5 feet thick and 8 feet wide in the center, expanding to 12 feet wide on each end, giving it a strong, massive appearance. The alcove is impressively deep, running 39 feet back into the hillside as measured from the front edge of the lintel.

251

This nearly equals its reach of 44 feet at the entrance. Most of this depth (22 feet) occurs behind the present waterfall and may have formed after the stream broke through the skylight. A steady seep from a crevice extending 6 feet into the ceiling 13 feet behind the waterfall illustrates how the natural bridge formed. If continued erosion widens the crevice into a skylight, there eventually could be two natural bridges here.

The floor of the alcove is only a few feet above the level of the Ohio River whose low bank is just beyond the road in front of the natural bridge. Its proximity to both the rich waters of the Ohio and to upland hunting areas made the alcove attractive to at least one band of aboriginal hunters. A small amount of bone, flint chips and other evidence of human occupation was excavated by archaeologists from a shallow pit still visible in the floor. These remains indicate that it was used as a short-term campsite (Murphy, 1975). Perhaps the water falling through its roof made the shelter too damp for long-term use. At present, this is the only natural arch in Ohio associated with prehistoric occupation.

It apparently played a part in some exciting historic events as well. In 1776, a group of Kentuckians including Robert Patterson, the founder of Lexington, traveled upstream on the Ohio River, intending to purchase supplies at Pittsburgh. Having camped across from "Hockhocking Island," the men ate their supper and stretched out by the fire. During the night, a band of Indians attacked the camp, killing two of the men and wounding all but two of the others. The next morning the survivors sought shelter in a hidden ravine, but constant rain caused them to move downriver to a "rock projection" where they would be better protected (Figure 143). One of the unwounded men left for Pittsburgh to get help, leaving the others to subsist on pawpaws and wild grapes until he returned nine days later with men and boats. Recent research indicates that the rock which sheltered Patterson and the others was the alcove at Mustapha Natural Bridge (Baker and Watkins, 2005).

Patterson recovered from his wounds and participated in several battles against the Indians and British, including George Rogers Clark's Illinois campaign, the Battle of Blue Licks and St. Clair's

Defeat. He became one of the three men responsible for founding the town of Losantiville, later renamed Cincinnati. Late in life he moved to Dayton, Ohio, and became a successful businessman whose investments included present Clifton Mill in the village of Clifton and the Patterson Mill, located in what is now Clifton Gorge State Nature Preserve. His business acumen passed down through the family, and his grandson John Patterson founded the National Cash Register Company (NCR) on the family farm in Dayton.

Ladd Natural Bridge (OH-A-WAS-02)* — Washington County
Span: 40 feet (12.2 meters); Clearance: 17 feet (5.2 meters)
Ladd Natural Bridge (figures 144–146; Plate 14), also known as Big Natural Bridge and one of the largest in Ohio, has the added distinction of being the only one with a history of having been a commercial tourist attraction. A faded sign that once appeared on an old barn along State Route 144 advised travelers to "See the Bridge"

Figure 144. Ladd (Big) Natural Bridge. The arch as seen from below. The skylight defining this arch is visible in the upper left corner of the photograph.

Figure 145. Ladd Natural Bridge. A roadside advertisement encouraging travelers to "See the Bridge."

(Figure 145). The Ladd family, long-time owners of the natural bridge, graded a road to it from their farmstead sometime between World Wars I and II. The entry fee of ten cents per person was low enough to entice many picnickers to visit the site for a pleasant day in the country. Some of the braver visitors even drove their cars across the bridge, making Ladd one of only two natural bridges in Ohio known to have carried vehicular traffic, the other being Cats Den Natural Bridge in Geauga County.

It would take a steady hand to guide an automobile across the deck of this natural bridge (Figure 146). At 12 feet wide and 5 feet thick, the lintel is strong enough to hold a vehicle and blocky enough in appearance to inspire confidence. However, on one side is a straight fall 17 feet to the floor of the skylight and on the other an equally precipitous drop 52 feet to the bottom of the cliff in which the bridge is carved.

It is this last drop which gives this natural bridge its impressive appearance. The bridge has formed where a small, unnamed stream plunges over a ledge into the valley of Gilbert Creek. Undercutting of the sandstone bedrock with subsequent retreat of the waterfall has created a cliff at this point. It is not a perpendicular wall of rock, however. There are three main layers of resistant sandstone 5 to 8 feet

Figure 146. Ladd Natural Bridge. Would you drive across it?

thick, separated from each other by layers of less-resistant, thin-bedded sandstone and shale. These weak layers have weathered back, leaving the thick sandstone layers jutting out like shelves. The highest of these sandstone ledges forms the lintel of the natural bridge. The middle resistant layer forms the floor of the skylight which extends only as far as the back face of the lintel above, leaving the bridge itself suspended over thin air. The tiny stream which plunges into the skylight cascades over the bottom two sandstone layers, adding the beauty of falling water to the scene when there is water available to fall.

Not all of the water flows over the falls, however. Some of it seeps down through crevices in the sandstone bed of the creek upstream from the skylight and joins other groundwater flowing along the bedding planes of the fairly impervious shale and out the face of the cliff, thereby illustrating how the natural bridge formed. Weathering of the bedding planes resulted in an alcove much like those now forming beneath all three layers of resistant sandstone. Widening of a crevice running parallel to the cliff face just behind the lip of the waterfall pirated the flow of the stream and sent it down through the roof of the upper alcove and out through its opening in the face of the cliff. The front edge of the alcove roof remains as the natural bridge.

255

This same process can be seen in action on the lower levels of the cliff. The present natural bridge will not be the last to form here. It is almost certainly not the first. Several large, angular slump blocks lie in a line across the valley at the base of the cliff. They appear to be the remains of a previous natural bridge formed in the middle sandstone layer directly beneath the present bridge. One large rock marking the junction of the lintel with the abutment is still in place and indicates that this older bridge was similar to Ladd in size and shape, having a flat top and an upward-arched bottom. Nowhere else in Ohio is the evidence of a collapsed natural bridge so strikingly evident.

Ladd Natural Bridge has also been called Big Natural Bridge in reference to the visual contrast between it and Little Natural Bridge a few miles away. Interestingly enough, Little Natural Bridge, described below, is the longer of the two and has a wider skylight. However, it has a narrower lintel which thins toward the front, giving it a less massive appearance. Nor does it stand so impressively high above its valley. Ladd Natural Bridge may not be the larger of the two in measure, but it is larger in grandeur.

The surroundings of Ladd Natural Bridge hold their own interest. A mature beech-maple-hemlock forest with a lush understory including Maidenhair Spleenwort (*Asplenium trichomanes*), Walking Fern (*Camptosorus rhizophyllus*) and Mountain Laurel (*Kalmia latifolia*) survives in the ravine below the arch. The drier upland forest is dominated by various species of oak. Rock outcrops and alcoves are common. The entire area is an outstanding natural treasure, which was why the Ladd family, who still owns the property, dedicated 35 acres surrounding the natural bridge as a state nature preserve in 1984. As such, it is accessible to the public with written permission from Ohio DNAP.

Little Natural Bridge (OH-A-WAS-03) — Washington County
Span: 44 feet (13.4 meters); Clearance: 26 feet (7.9 meters)
Located just a few feet from a road, Little Natural Bridge (figures 147 and 148) has been especially vulnerable to unappreciative visitors over the years. In spite of the graffiti and trash that has accumulated at the site, it remains one of Washington County's most impressive scenic features. Formed in a layer of resistant Permian

Figure 147. Little Natural Bridge. The road is just beyond the back edge of the skylight, which opens behind the arch in this view.

sandstone, Little Natural Bridge spans a tiny, unnamed tributary of Burnett Run. As usual, the architect of this feature is a stream of very small dimensions and sporadic flow, having a watershed of only four or five acres.

The lintel of this arch is only 8 feet wide. This, combined with a pronounced thinning from back to front and the very gentle outward curve of its front edge, gives this natural bridge a particularly graceful appearance (Figure 148). This is further emphasized by the large width of its skylight (14 feet) which gives the illusion of a span of rock floating on air. The bottom of the lintel displays extensive bas-relief filigree resulting from the erosion of less-resistant sandstone from around knots of rock more strongly cemented with iron compounds.

Little Natural Bridge retains its impressiveness in spite of the scrubby appearance of its cut-over surroundings and the presence of a road culvert which now carries the tiny unnamed stream into its skylight. Like an elegant matron fallen on hard times, this important natural feature deserves better.

Figure 148. Little Natural Bridge. A view from underneath. The skylight is on the right.

Irish Run Natural Bridge (OH-A-WAS-04)* — Washington County
Span: 22 feet (6.7 meters); Clearance: 5 feet (1.5 meters)
Irish Run Natural Bridge (figures 149 and 150) is a complex structure difficult to pin down with figures and, as a result, the dimensions as given for this feature are a bit deceiving. Although it is basically a breached alcove cut into Permian sandstone and thus similar to the other natural bridges in Washington County, it has had a slightly different erosional history.

This is a massive natural bridge. The blocky lintel is 18 feet wide and 6 feet thick. Its back edge exhibits the flat face typical of joint control and most likely represents the front edge of the crevice through which the intermittent stream now dropping into the skylight was first captured and sent through the roof of the alcove. This face

258

Figure 149. Irish Run Natural Bridge. A view from beneath the arch.

slants toward the front of the lintel as it descends. If it does indeed represent one side of the original crevice, this slant would have helped channel water beneath the resistant layer of rock forming the arch.

The skylight which resulted from the widening of this crevice is 22 feet long and 26 feet wide perpendicular to the lintel. Its angularity shows further evidence of the influence of joints. In fact, the two joints which were most important in forming the skylight can be traced into the wall of the alcove. One of them continues the line of the back side of the lintel and the other continues the back wall of the skylight. Both parallel the face of the cliff in which the arch is formed and may represent stress relief fractures. Collapse of the alcove roof between these two joints created the somewhat rectangular skylight. Several large blocks of sandstone littering the floor of the alcove make this one of the few natural bridges in Ohio where evidence of such collapse is still in place.

The length of the skylight represents the length of the natural bridge as originally formed. The 5-foot clearance reported was measured near the center of the opening beneath the lintel. The alcove, however, extends under cover another 17 feet behind the southern abutment of the natural bridge, and the floor drops 20 feet below its

Figure 150. Irish Run Natural Bridge. A view through the entrance from inside. The stream is flowing out beneath the lintel at the lower right corner. The enlarged alcove is to the right.

level at the skylight. Unlike streams at similar natural bridges which usually flow straight out beneath the lintel after plunging through the roof of the alcove, this stream meandered to the south before curving back to resume its original direction. In the process, it cut away the side of the alcove as it deepened its bed. This enlarged part of the alcove resembles a partial dome. The inside of the stream-cut curve is a classic slip-off slope, a rounded slope formed on the inside of the bend of a stream which is both deepening its bed and enlarging the bend. Viewed from the inside, the entrance to the recess resembles a wide slot sloping down to the south at a 45-degree angle (Figure 150). The skylight and the natural bridge it defines are in the upper end of the slot. The stream exits the enlarged portion of the alcove at the lower end of the slot.

Irish Run Natural Bridge, named for the small stream flowing through the valley below it, shows why Ohio's natural bridges have not received much attention. Although impressively large for Ohio, it is surrounded by a thick second-growth woodland which makes it difficult to see and almost impossible to photograph, illustrating one of the disadvantages of pursuing geology in the middle of the Eastern Deciduous Forest. Fortunately, this interesting feature and nearby Fern Pillar are both part of Wayne National Forest and accessible to the public, although not easily so. A good hike is required to reach them, but once there, the sight is well worth the effort.

Arches in Quaternary Strata

Ohio's final emergence above sea level occurred 243 million years ago during the Permian Period and inaugurated a time during which erosion rather than deposition became the dominant process affecting the surface geology of the state. With the advent of the Pleistocene glaciations, sedimentary deposits once again formed on the surface of Ohio. These strata consist almost entirely of rock rubble transported varying distances from its original location by the advancing glaciers and then deposited when the ice stopped expanding and melted back. The rubble was battered and ground down during its transport by the ice and flowing water, and now consists of everything from the fine particles of clay to boulders the size of small cars, sometimes segregated according to size by meltwater streams, but more often mixed together in heterogeneous deposits called till. This material is usually unconsolidated, found as the glacier left it. The one important exception however, is cemented glacial gravels.

Unlike most sedimentary rock which appears to have taken eons to solidify, cemented glacial gravel, which can be considered a form of conglomerate, formed over the course of a few hundred or thousand years. It consists of glacial gravels containing everything from coarse sand to sizeable cobbles cemented together with calcium carbonate. Groundwater highly charged with lime, most likely dissolved from calcareous rocks in the gravel itself, was the agent responsible for deposition of the cement. It is not unusual to find individual water-worn glacial cobbles in calcareous till with a coating of calcium carbonate. Sometimes the entire surface of the rock is covered by such deposits, but more often they are discontinuous. Now and then two or three cobbles will be found cemented together. Rarely, a lens of cemented glacial gravel forming a conglomerate will be found. Such deposits are usually small, covering an area measured in hundreds of square feet rather than miles. While small in extent and young in age, such glacial conglomerates can still give rise to natural arches if the conditions required for arch formation are present. So far, two examples, both small, have been found in Ohio, and one of these unfortunately appears to have been destroyed.

David Road Natural Bridge (OH-A-MON-01) (apparently destroyed) — Montgomery County
Approximate Span: 4.5 feet (1.4 meters); Approximate Clearance: 3 feet (0.9 meter)

David Road Natural Bridge (Figure 151) is known only from an illustration found in A. F. Foerste's *Geology of Dayton and Vicinity* published in 1915. A sparse caption reading "Natural bridge along road south of David church" along with the illustration's position at the head of a chapter entitled "General Observations of the Gravel Areas South of Dayton" offered a few vague clues as to where this feature might have been located. A David Road was noted in the region discussed in the chapter and there was even a church on it. My initial reconnaissance trip, however, showed that much had changed

Figure 151. David Road Natural Bridge (Foerste, 1915).

262

in the century since the photograph of the arch was taken. The city of Dayton had overtaken the area and what was once open countryside had become a part of crowded suburbia. Almost by chance, an outcropping of cemented glacial gravel nearly hidden by a dense growth of young trees was sighted along the roadside in someone's front yard. It appeared to be the north abutment of the bridge as shown in the photograph. Unfortunately, road widening and land grading over the past century have greatly changed the contours of the area. If this outcropping is indeed the abutment of the bridge, then the bridge itself is gone, no doubt knocked down by heavy equipment in the quest for "improvement." No other such outcroppings were found along the road.

In the Foerste photograph, a young man is seen sitting on the left abutment of the natural bridge and his inclusion made it possible to arrive at approximate measurements of the span and clearance. The lintel appears to have been barely a foot thick at its narrowest point. The photograph also shows the coarse, pebble fabric of the rock surrounding the opening of the arch and the rough, weathered-out layering expected in exposed beds of cemented, water-laid gravel. The arch crossed the small valley of an apparently intermittent stream which flowed toward the photographer, so it was indeed a natural bridge. Although it is difficult to say for certain how the arch formed, it was most likely a breeched alcove.

Cobble Arch (OH-A-MIA-02) — Miami County
Span: 3.6 feet (1.1 meters); Clearance: 1.2 feet (0.4 meter)
Having no desire to end this catalog of Ohio's natural arches on a depressing note, I was happy to find Cobble Arch (Figure 152; Plate 15) to offset somewhat the loss of David Road Natural Bridge. Although it is nearly off the bottom end of the significance scale even for Ohio, this small arch is very important as it is the only known existing natural arch formed in glacial conglomerate in the state. It is also more impressive than its measurements indicate since they represent only the size of its skylight. The front opening of this breached-alcove arch is 9 feet wide and 6 feet high. While not large in comparison to the major alcove-type arches of southeastern Ohio, it has a surprising presence in the miniaturized topography found here near the junction of

Figure 152. Cobble Arch. A view of the entrance above the unnamed stream. The skylight is visible at the rear of the alcove. Note the large cobbles which form the bulk of this conglomerate, and the apparent bedding visible near the bottom of the left abutment.

the small unnamed stream below it with the Stillwater River. This is, after all, western Ohio, and any topographic relief should be appreciated.

The glacial conglomerate in which the arch is formed is made of coarse sand, gravel and rounded cobbles, one of the largest of which measures 12 inches by 6 inches. The arch was named for these striking rocks. Glacial erratics of metamorphic and igneous origin carried south from Canada are abundant, as are calcareous rocks from more local sources. Since these calcareous rocks do not appear to have lost much of their calcium carbonate content to solution by groundwater, the conglomerate is assumed to be of Wisconsinan age. It is a curious juxtaposition to find Cobble Arch, formed in one of Ohio's youngest geological deposits, less than 2 miles from Greenville Falls Arch which formed in Silurian dolomite, some of the state's oldest exposed bedrock.

This bed of conglomerate appears to be of small extent. Outcrops were traced just 80 feet upstream and 30 feet downstream from the arch on the east side of the Stillwater River, and 120 feet upstream on the unnamed tributary fronting the arch. The banks of both streams have been covered in places by recent fill which may hide further outcrops. Part of this fill takes the form of old sidewalk slabs on the banks and in the bed of the unnamed stream. Interestingly, the same type of cobbles found in the glacial conglomerate were used to make the cement slabs which might easily be mistaken for chunks of the conglomerate itself were it not for their rectangular shape and flat top surfaces. While uncommon, cemented glacial gravel conglomerates do exist elsewhere in Miami County. One found near Piqua had a reported thickness of 40 feet (Hussey, 1878).

Cobble Arch is found in a vertical outcrop rising about 10 feet above the bed of the unnamed stream very near its junction with the Stillwater River. Differential erosion has emphasized otherwise obscure horizontal bedding planes and carved two fairly deep alcoves side-by-side. The arch is in the larger, western alcove which is 10.7 feet deep. The skylight forming the arch is located at the rear of the alcove near its western side. This narrow opening is a bit less than 4 feet long and is found at the bottom of a sinkhole 2.2 feet deep, formed in clay overlying the conglomerate. The sinkhole is oddly sited on the crest of a narrow ridge which drops steeply from the upland to the junction of the Stillwater River and the unnamed stream. This ridge is not much wider than the sinkhole itself and has a rounded profile that would tend to send surface water down its sides rather than along its narrow top and into the sinkhole. Even so, enough sheet wash evidently finds its way into this funnel-shaped hole to build a continuous fan of clay and gravel extending from the bottom of the skylight across the floor of the arch to the edge of a 3-foot-high ledge dropping into the unnamed stream. Perhaps the sinkhole formed as the two streams were eroding their channels down to their present levels. If so, the clay and gravel fan could be a relic of a time when the narrow ridge was much wider and the drainage basin feeding the sinkhole was much larger. This would mean that the perfect centering of the sinkhole on the narrow ridge was a fortuitous accident, a conclusion difficult to accept. Here is a minor mystery that might be solved by a brave observer sitting in the alcove during a heavy rainstorm and noting whether or not surface water still makes its way down through the sinkhole.

Black soot on the ceiling of the alcove shows that such an adventure might already have occurred. Of greater concern than the arch's use as a neighborhood shelter for adventurous youths is its possible damage or destruction by uninformed construction activities. An impressive brick Italianate home built in the mid-to-late 1800s sits on the upland above it. A more recent building of some official character stands on the opposite side of the unnamed stream which was encroached upon by fill during the building's construction. A thick growth of bush honeysuckle now obscures the entire valley of the small stream, which may or may not serve to protect this unusual and important geological feature.

Potential Additions to the Catalog of Ohio's Natural Arches

The features listed below have not yet been surveyed to determine whether or not they meet the qualifications for listing as natural arches. Time constraints, difficulty of access or lack of adequate information have so far prevented their documentation. Some of them have been viewed, but could not be measured. Others are known only from photographs, literature sources or personal communications.

Merkles Natural Tunnel — Geauga County

According to two articles in *Pholeos* (Hobbs, 1982; Luther, 1988) reporting on joint fracture "caves" found in Munson Township, Geauga County, two natural bridges along with many caves and a box canyon are located in what are called Darts and Merkles caves. From their description, the features have formed in the Sharon Formation. Luther (1988) reports that Merkles Cave has the appearance of an unroofed limestone cave, even though it has formed in conglomerate. However, approximately 30 feet of the chamber retains its roof to form a "true natural tunnel." This feature, privately owned, has not yet been documented.

Eagle Rock Arch — Hocking County

Across the narrow gorge of Old Mans Creek from the Old Mans Cave Visitors Center is Eagle Rock (Figure 153), a long projection of

266

Figure 153. A potential pillared-alcove-type arch beneath Eagle Rock in the gorge of Old Mans Creek, Hocking Hills State Park.

stone jutting out from the rim of the cliff above the small stream below. To more imaginative visitors it appears to be the head of a gigantic eagle and in this it resembles Raven Rock projecting out over the much wider valley of the Ohio River in Scioto County. Beneath it lie two long indentations separated by a stocky pillar. The only view of this feature is from below and from this vantage point it is difficult to be certain that the alcoves connect behind the pillar, but they appear to do so, making this potentially the second example of a pillared-alcove-type arch found within the upper part of the gorge of Old Mans Creek, Old Mans Pantry being the other. It appears to have a span of about 5 feet and an approximate clearance of 3 feet. Although the alcoves appear to be inaccessible without the use of technical climbing equipment, at least one intrepid — or more accurately, foolhardy — person has managed the feat as is proved by an inscription on the outside face of the pillar. Not being foolhardy, my measurements are approximations made from ground level.

Canters Cave Natural Bridge — Jackson County

The Elizabeth L. Evans Outdoor Education Center Canters Cave 4-H Camp Inc., in Jackson Township, Jackson County, contains several rock-bound hollows. A field trip to the site many years ago took the group I was part of through several of these. In one, a small intermittent stream was falling through a hole in the roof of a sizeable alcove at the head of the hollow, forming a small natural bridge. At the time I did not have measuring equipment with me, but the skylight appeared to be at the lower end of the significance scale for Ohio.

Two Arches — Lawrence County

Two arches have been reported in Fayette Township, Lawrence County, north of US Route 52. One is called a "double arch."

Montgomery County Arch — Montgomery County

James A. Foster of Miami Township, Montgomery County, reported a small arch with an opening approximately 1 foot high on the east side of the Dayton-Cincinnati Pike (J. D. Foster, personal communication). Recent attempts to relocate it failed. Photographs of the arch appear to indicate that it has formed in consolidated glacial gravel. If so, and if it proves to be large enough to qualify for the state list, it would be an exciting addition.

Chapter 4

About Natural Pillars

Natural rock pillars are the neglected step-child of geology. A quick review of a sample of geological literature indicates that, when referred to at all, pillars usually are given only passing notice. Even when treated in more detail, the accounts are brief and usually mention them as illustrations of geological principles rather than as features of interest in their own right. This tendency on the part of geologists to ignore natural pillars may be due in part to the obvious simplicity of their formation. They are "erosion remnants," rocks left behind by the weathering processes that removed the bedrock originally surrounding them (Figure 154). Beyond that, there seems to be little to say.

Popular interest in natural rock pillars is slightly higher than that of geologists, as evidenced by the large number of such formations which have been given names. Pillars also appear at infrequent intervals in newspaper and magazine articles, although more as natural curiosities and tourist attractions than as features of serious inquiry.

Given the peripheral status of pillars, it should be no surprise that catalogs of such formations are extremely rare. They are usually included in lists of geological features of interest to state natural resource agencies and a few have become centerpieces of parks and nature preserves acquired to protect them. Such attention is unusual, however. If a natural rock pillar finds itself in a protected status, it is usually because it has had the good fortune to sit on a piece of property acquired to protect some other natural or recreational resource. In the American West where truly spectacular examples can be found

Figure 154. Big Pine Creek Pillar, Hocking County — just an "erosional remnant."

by the thousands, they are too numerous to count, let alone catalog. In the humid East, however, they are thought to be too rare to be considered worthy of anything more than a cursory glance.

The history of this survey is itself a good example of the usual attention given to natural rock pillars in geological research. Begun as a study of natural arches in the state of Ohio, it grew to embrace natural pillars only when they insisted on intruding themselves into the process. The town of Mineral is responsible for that. To the casual observer, the village may not appear to have much going for it, but it does have two important geological features of great interest: Mineral Arch to the east and Mineral Tea Table to the north. Having surveyed the arch, it seemed only reasonable to at least look at the tea table since it was nearby. After watching it appear to rise out of the earth during the steep climb to its hilltop location, I could no longer ignore the casual references to such features that continually appeared during the course of my research.

While the following discussion may appear to be much ado about a little bit of nothing, it should be understood that the lack of attention given to these features means that the basic groundwork for their study has yet to be laid. This survey is an attempt to do just that, although it should be emphasized that the definitions and terms adopted here are meant to apply only to those features found in Ohio. Researchers in other areas are welcome to them if they find them useful. It is hoped that the inclusion of natural pillars in this survey will help to bring these interesting features out of the shadows and give them their proper place in the study of Ohio's natural history. Their importance is indicated by the fact that 18 natural pillars are listed in the catalog and an additional 14 features which have not yet been documented are described as potential pillars. Like natural arches, natural pillars form a larger part of the state's scenery than has previously been recognized.

Definitions: What is a Natural Pillar?

Defining natural pillars is even more difficult than defining natural arches. A natural pillar is usually considered to be an elongated, upright remnant of rock, but these can be found in an infinite variety of shapes and sizes. The famous pinnacles of Bryce Canyon in Utah (Figure 155) would certainly qualify in the minds of most observers, but is Wyoming's equally famous Devils Tower (Figure 156) also a natural pillar? The confusion extends to what they should be called as well. A quick survey of geologic texts shows them referred to as mushroom rocks, hoodoos, sea stacks, pedestal rocks, towers, needles and pinnacles, all names which are used to discuss specific types of vertical rocks. Popular usage presents its own list of names: chimney, chimney rock, tea table, standing rock. The following discussion is an attempt to bring a bit of order out of this profusion.

Figure 155. A pinnacle in Bryce Canyon National Park, Utah.

The term "pillar," defined in the American Geological Institute's *Glossary of Geology* (1960) as "A column of rock remaining after solution of the surrounding rock," has achieved a measure of official sanction and appears to be the appropriate generic term under which any others found useful can be included. It will, however, be further refined for use in this report. Since processes other than solution can be involved in pillar formation, that word in the original definition will be replaced by the more general term "erosion" which includes the process of solution as well as mechanical weathering and transportation of debris.

According to dictionary definitions of "pillar" and "column," words often used to define each other, the primary criteria are structures

272

Figure 156. Devils Tower, Wyoming. A natural pillar?

that are slender and vertical. Originating in the realm of architecture, pillars and columns are primarily supports for some sort of super-structure, but use of the terms has expanded to include such struc-tures standing alone as monuments to people or events. From that use, the terms have broadened further to include anything resembling such monuments in form or function, which is where natural forma-tions of similar shape can be included. Since columns are generally considered to be cylindrical in cross-section, the more general term "pillar" is to be preferred in the geologic context. A more correct term for our purpose would be "natural pillar," indicating one formed by naturally occurring processes; when the word "pillar" appears alone in this report, the word "natural" is to be presumed.

In discussing Ohio's natural pillars, a few additional definitions will be useful (Figure 157). The point at which the pillar attaches to the underlying bedrock is called its "base." Its highest point above the base is its "top." A pillar's height will be the greatest vertical dimension measured between the base and the top. The shortest horizontal dimension found at the narrowest point closest to its base will be its "width." The longest dimension measured at the same point will be considered the pillar's length. Natural pillars often swell at the base into a broader expanse of bedrock rising above the general level; this expanse will be called the "platform."

Natural pillars are erosion remnants of native bedrock. They are part of that bedrock and still connected to it, but only at their base. This removes slump blocks, glacial erratics and other transported rocks which may happen to be slender and vertical, but only because they were moved into a position which causes them to resemble true natural pillars. In conformity with the terms used for natural arches, such features are considered to be gravity pillars. Natural pillars may

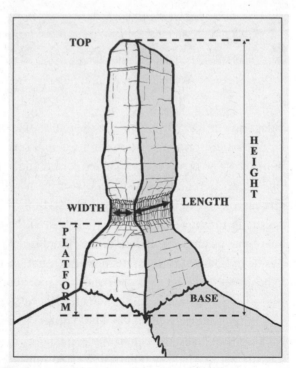

Figure 157. Terms used in describing natural pillars.

experience movement along bedding planes, either in the body of the pillar itself or in its underlying bedrock support, without losing their status as true natural pillars.

It is also necessary to establish just what "slender" means. There is great variety among pillar-like forms and devising a definition broad enough to include those features which most observers would instinctively consider to be natural pillars, but narrow enough to be meaningful, is not an easy matter. Since the most distinctive feature of a pillar is the ratio between its height and its width, this quality was made the core of the definition and it was generously interpreted in order to include the widest selection of examples. Therefore, a vertical erosion remnant whose height equals or exceeds twice its narrowest horizontal dimension, its width, is considered to be a natural pillar.

During the course of the survey a number of striking features were found which appear to be pillars from one side, but which are much broader when viewed from another angle, looking more like walls or upright slabs. Although slender in one horizontal dimension, they do not resemble columns. There is, however, no definite point at which a pillar becomes a wall and so it was thought best to allow a little subjectivity to intrude. Thus the definition of "natural pillar" used herein will be, "A narrow, vertical erosion remnant of bedrock remaining attached at the base, whose height is equal to or greater than two times its width and whose greatest length measured at the same point is less than the height, which must equal or exceed 6 feet."

Terms Used in Describing Natural Pillars

With a definition of the generic term "pillar" now established, we can look at several other descriptive terms which have found a place in the literature on the subject (Table 3).

The picturesque word "hoodoo" consistently refers to rock pillars of grotesque shape eroded from horizontal strata of varying resistance in arid regions where rainfall is infrequent and of short duration. Although Ohio has its share of hoodoo-like natural pillars, it is not an arid region and the term is not applicable here.

"Sea stacks," remnants of rocky headlands left behind by retreating coastlines, appear to have received more attention in the literature

Table 3. Definitions of Natural Pillars

Pillar	A narrow, vertical erosion remnant of bed rock remaining attached at the base whose height is equal to or greater than two times the narrowest horizontal dimension and whose greatest length measured at the same point is less than the height which must equal or exceed 6 feet.
Chimney	A type of natural pillar characterized by generally vertical sides with horizontal dimensions (width and length) that are roughly equal.
Tea Table	A type of natural pillar with a narrow lower section and a wider top, similar in form to a one-legged table.

than any other form of natural pillar. They are defined as a "Small steep-sided rocky projection above sea level near a coast" (Howell, 1969). They appear in infinite variety from truly slender columns to more massive, rounded forms resembling the hay stacks from which they apparently took their name. Given the dynamic, high-energy environment in which they were formed, sea stacks are geologically short-lived features. In Ohio, sea stacks, which perhaps more properly should be called "lake stacks," are found only along the Lake Erie shore on Marblehead Peninsula and the Erie Islands where bedrock forms the coastline. Although there are historical records of Ohio sea stacks, and potential stacks are presently being formed, no true sea or lake stack is now known to exist within Ohio.

The popular term "chimney rock" has also found a place in the literature as well as in dictionaries (e.g., Guralnik, 1976) where it has been defined as "A chimney-shaped body of rock rising above its surroundings, or partly isolated on the face of a steep slope." The most famous example of this category of pillar is Chimney Rock, a noted landmark in Nebraska along the old Oregon Trail (Figure 158). There are numerous pillars named Chimney Rock east of the Mississippi River, indicating that the name of the Nebraska example was imported by immigrants moving west. Even Ohio has a Chimney Rock located in Pike County. Given the term's eastern associations and established usage, it seems proper to retain it to denote a specific type

Figure 158. Chimney Rock along the Oregon Trail in Nebraska.

of natural pillar characterized by generally vertical sides and with horizontal dimensions (width and length) that are roughly equal.

Another popular term for pillars commonly found in use east of the Mississippi River is "tea table," which refers to natural pillars with a narrow lower section and a wider top, similar in form to a one-legged table. Although the term has been utilized only in popular literature, it will be adapted here to indicate this specific type of natural pillar. In describing tea tables, the term "leg" will be used for the thinner lower part, and the term "table" for the broader upper part (Figure 159).

Several pillars in eastern Ohio are referred to as "standing stones." While descriptive, this term has unfortunately been applied to a broad range of geological features. In fact, the most famous of Ohio's standing stones is not a pillar at all, but Mount Pleasant, a 200-foot-high hill historically called Standing Stone because of its sheer face which now overlooks the Fairfield County fairgrounds in Lancaster (Figure 160). In this report, the term "standing stone" will be reserved for features which almost, but not quite, meet the stricter definition. In most such cases, these "almost pillars" do not have the required relationship between height, length and width that would allow them to be listed. A few of the more interesting of these are included in the list of potential natural pillars at the end of Chapter 5 — "A Catalog of Ohio's Natural Pillars."

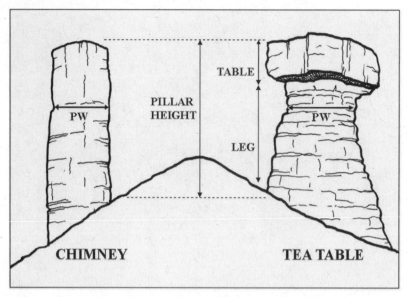

Figure 159. Chimney rocks and tea tables. Shown here are terminology and measurements used to describe these features. PW = pillar width.

Figure 160. Mount Pleasant in Rising Park, Lancaster, Ohio. This distinctive hill is also known as Standing Stone.

278

Measuring Ohio's Natural Pillars

Measuring natural pillars is less complex than measuring natural arches. Overall size can be indicated most readily by measurements of height and width, and these two measurements are given in the headings before each pillar description. There are, however, a few complications to the systematic determination of reference points for making measurements, the most frustrating of which is the frequent absence of definite breaks between units. Pillars often swell gradually onto their platforms with no definite separation point. As a rule, the height of the platform is included in the height of the natural pillar. The legs of tea tables can likewise broaden into their overhanging tables with no perceptible break. Determining where one ends and the other begins is often an arbitrary decision, although confined within perceivable limits.

Many natural pillars are located on steep slopes and exhibit a large difference in height between the upslope and downslope sides. The most impressive view of a pillar is that encompassing its greatest rise which is nearly always found on the downslope side, but the truest expression of the extent of a pillar's separation from the surrounding bedrock is given by its smallest vertical dimension which is almost always found on the upslope side. This is the measurement given in the catalog as a natural pillar's official height. Its height on the downslope side is also recorded for descriptive purposes when it can be measured safely.

Pillar width is the measure of the shortest horizontal dimension at the narrowest point nearest the base. When this cross-section is noticeably oblong, its longest horizontal dimension is recorded as the length. Since even chimney rocks rarely have perfectly vertical sides, the profiles of natural pillars usually exhibit a number of narrowed sections. Utilizing the narrowest one closest to the base gives the clearest impression of the actual support of the pillar. Determining which narrow section is nearest the base is usually not difficult, but in some cases a frankly arbitrary decision is again required.

For tea tables, height is measured from the base to the top of the feature and includes both the leg and the table. Width and length are still measured at the narrowest point nearest the base, and will as a result describe the leg. For descriptive purposes, the thickness of the table as well as its length (longest horizontal dimension) and width (shortest horizontal dimension perpendicular to the length) may also be given. Where the height of the leg alone is not given, it can be deduced by subtracting the table thickness from the overall height of the tea table. It should be noted that measurements, especially of tables, are often difficult or impossible to obtain. In such cases the needed dimensions are estimated.

The Question of Significance

As in the case of natural arches, the significance of natural pillars is related to their size. There is an infinite number of bumps and protrusions on the bedrock landscape that fulfill the definition of natural pillar, but the small size of many of these features renders them insignificant from a human viewpoint. Rarity is also a determiner of significance. In the canyon country of the American West, thin spires of stone 200 and even 300 feet high are not uncommon. In fact, pillars are so very common in the region that they are given little consideration at all unless they are very large or very oddly shaped. Geology and climate have converged in Ohio to make pillars not only much smaller, but also much less common, than is the case in arid regions of the West. While their diminutive size decreases their significance, their rarity increases it enough to make them a subject of interest. For the purposes of this report, a pillar is considered significant in Ohio if its height equals or exceeds 6 feet, a value observers in less geologically challenged regions of the country might think laughable. In the flatter Midwest, however, it allows a list of the more interesting and informative pillars to be compiled.

A Brief Consideration of Pillar Names

A surprising number of natural pillars have been named. Unfortunately, the authors of these names were often less than creative. The United States has no lack of "Chimney Rocks." A quick look through the Geographic Names Information System maintained by the US Geological Survey shows that, in addition to the Ohio and Nebraska examples already mentioned, Alabama, Arizona, Colorado, Illinois, Kentucky, New Mexico, Utah and West Virginia are blessed with one or more versions. Wyoming has nine. If modifications such as "The Chimney" are included, the count is even higher. One can only conclude that Americans have been woefully unimaginative in bestowing names. Evidently it was easier to apply a name already familiar from previous usage to similar features found as the population moved into new territory. Only when explorers and settlers found themselves facing the countless pillars of the western canyon lands were they forced to be more creative; but even here where names like Adam and Eve, Rabbit Ears and Candlestick Tower found a place, the ubiquitous Chimney Rock still managed to hold its own.

Names including "tea table," on the other hand, seem to be an eastern institution; as one reviewer of this manuscript said, "It's hard to imagine cowboys sitting around a tea table." For some reason, tea tables are often associated with Satan. New Jersey and Indiana both have a Devils Tea Table, and no less a personage than Mark Twain himself noted one on the Mississippi River. Ohio has several, including one in Ohio Caverns. In at least two cases, in Indiana and Ohio, the devil was given credit for these unusual rock formations because local folklore attached tales of mystic Indian rites to them, reflecting a recognition of the Native American belief that unusual places in the landscape were possessed of great spiritual power.

All natural pillars included in this survey were named. Where possible, names already in use were retained. In the case of multiple chimneys and devils tea tables, the names were modified in order to distinguish between the features. Where no names existed, names were bestowed by the owner, land manager or the author of this report. In such cases, name-givers were encouraged to be more creative than their predecessors.

How Natural Pillars Form

The same weathering processes which form arches also form pillars. In most cases, the enlargement of vertical crevices isolates an incipient pillar from a cliff face (Figure 161). As weathering progresses, the crevice widens, increasing the distance between the pillar and its parent bedrock exposure. At the same time, the pillar itself is continually thinned by weathering which attacks it from

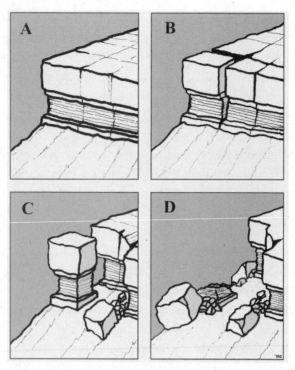

Figure 161. The life cycle of a natural pillar. **A:** A projection forms in a bedrock cliff broken by vertical joints. **B:** Widening of the vertical joints and cliff retreat isolates a block of bedrock. **C:** Further joint enlargement and cliff retreat completely isolates the block, resulting in a natural pillar. Erosion weathers the pillar on all sides. If the bedrock is composed of a massive overlying caprock underlain by weaker rock which erodes faster, a tea table is formed. If the entire height of the cliff is made of massive rock, a chimney will develop. **D:** Continued erosion weakens the pillar which eventually becomes unstable and collapses.

every side until it reaches the point of instability and collapses. This thinning appears to take place at a slower rate than does the retreat of the cliff which birthed the pillar, as is shown by separations which are often wider than the pillar itself. As the vertical crevice between the cliff and the pillar is widened, the flow of groundwater through the pillar is cut off, eliminating one potent source of erosion. This enables the isolated pillar to withstand weathering longer than the groundwater-soaked cliff behind it in much the same way as happens in isolated natural arches.

One of the most common sites of pillar formation is the junction of two valleys whose rims are composed of a cliff-forming layer of resistant rock. Should the acute angle formed by the cliff at the junction be separated from the bedrock farther back by a vertical fracture, it may be isolated as the crevice is widened by weathering and the cliff retreats. The presence of intersecting valleys is not a requirement, however. Any vertical point of rock projecting from a cliff can become a pillar if a means of separating it from the parent bedrock exists. In some cases, two or three intersecting vertical crevices may be widened into a network of passageways among isolated bedrock remnants creating a feature sometimes referred to as a "rock city." If weathering reduces any of these remnants to a narrow, vertical form, then a natural pillar results.

If the cliff is composed of homogeneous rock, the pillar carved from it will also be composed of homogeneous rock and the vertical face of the cliff will be expressed in it as vertical sides. The resulting pillar is a chimney. If, on the other hand, the cliff is made of a layer of resistant rock capping a layer of less-resistant rock, the pillar carved from it will be similarly composed. The same pattern of undercut base topped by an overhanging vertical wall found in the adjacent cliff will be seen in the pillar. In the pillar, however, undercutting is able to proceed on all sides. The result is a pillar with a base narrower than its top — a tea table.

The profile of tea tables formed in layered rocks is obviously due to differences in resistance of the layers, but tea tables are also found in massive or obscurely layered rock. These tea tables are usually the result of differences in the sedimentary grain or texture of the rock. As is the case with bedrock-texture natural arches, these pillars result from more-resistant knots of rock being exposed as the less-resistant matrix around them is weathered away, a process that can form chimneys as well as tea tables. Bedrock-texture pillars can be recognized by their pitted surfaces and lack of obvious layering. Tea tables formed in this fashion tend to have rounded profiles with the legs swelling into bulbous tables with little or no break defining the junction (Figure 166). Their tea table form may be emphasized by precipitation running down the sides of the pillar, a process referred to as the "teapot effect" (Scheidegger, 1970) because it resembles water running down the spout of a teapot instead of into the targeted cup. This concentrates moisture at the base of the pillar and encourages more rapid weathering there. Shade cast by the widening table as well as by surrounding vegetation helps to keep the base moist, extending the length of time chemical weathering can affect it. The top of the pillar, exposed to sun and wind, dries more quickly and weathers more slowly. Although this process might also occur in tea tables formed of layered rock, the profile of such features is usually more angular with a definite junction between the table and the leg. Once the top of the leg erodes back beneath the angular edge of the table, a drip line forms, allowing precipitation to fall directly to the ground rather than run down the leg.

Formation of Non-Listed Natural Pillars

Just as it is possible for arch-like structures to form that do not meet the definition of a natural arch, so it is possible for pillar-like structures to form that do not meet the definition of a natural pillar. Thin slabs of bedrock released by erosion from a cliff often assume an upright position during their migration downslope (Figure 162). Because these rocks are not attached to the bedrock, they are not considered to be true natural pillars. Bedding planes which tilt at a different angle or to a different degree from those seen in nearby bedrock outcroppings are a good indication that the suspected natural pillar is actually a gravity pillar. If the rock is lower on the slope than similar rock in nearby outcroppings, it almost certainly has migrated downward and so is unattached.

Pillar-like shapes can also form through deposition of mineral deposits. The most familiar example is the stalagmite, a true pillar in everything except manner of formation (Figure 163). Stalagmites form by the deposition of calcium carbonate on the floor of a cave by highly charged groundwater dripping from its ceiling. As each drop of water evaporates, it leaves behind its minute burden of dissolved material. Over time these tiny increments of reconstituted rock can build up

Figure 162. A slump block resembling a natural pillar. Although tall and slender, this rock, found in John Bryan State Park, is no longer connected to its parent bedrock and so is considered to be a gravity pillar.

Figure 163. A pillar-shaped stalagmite located in Ohio Caverns.

Figure 164. Liberty Cap, a pillar-like spring deposit in Yellowstone National Park.

into surprisingly large stone towers. The examples found in Ohio are, like her caves, small but picturesque.

An unusual example of a deposition pillar which formed in the open air is the Liberty Cap, a famous sight in the Mammoth Hot Springs area of Yellowstone National Park (Figure 164). This interesting pillar-like structure formed through mineral deposition around an isolated hot spring which managed to keep its surface even with the rising rim of deposited rock. Eventually the water level reached a point at which the underground forces pushing the water up could force it no higher and the spring was extinguished. Although Ohio has no hot springs, it does have many travertine mounds formed by the deposition of calcium carbonate around cold-water spring outlets, most notably that of the Yellow Spring in the village of Yellow Springs (Figure 165). None of them, however, have formed a pillar.

Figure 165. A calcium carbonate mound precipitated from water at Yellow Spring in Glen Helen.

Chapter 5

A Catalog of Ohio's
Natural Pillars

The following catalog is not intended to be a complete listing of natural pillars found in Ohio. It is simply a by-product of the survey of Ohio's natural arches and presents information discovered while I was pursuing the original topic. Where pillars were stumbled across during field trips or could be visited easily while investigating arches in the same area, they were surveyed and measurements were taken. In only a few cases was a natural pillar the primary object of a trip. To give a more complete idea of the possible importance these features have in the Ohio landscape, natural pillars which were not visited, but which could be verified through other means, also are included. These pillars, listed at the end of the catalog, were not assigned feature numbers. A complete list of cataloged pillars is presented in Appendix II.

Pillars in Ordovician Strata

No pillars have been reported from Ordovician strata in Ohio.

Pillars in Silurian Strata

Fort Hill Tea Table (OH-P-HIG-01) — Highland County
Height: 12 feet (3.7 meters); Width: 6 feet (1.8 meters)
Fort Hill Tea Table (Figure 166), located in the depths of Baker Fork Gorge and surrounded by one of Ohio's largest collections of natural arches, is one of the state's more unusual pillars. Most tea tables in Ohio have an angular profile, but this one is rounded in outline, resembling an automobile headlight of the 1930s.

Figure 166. Fort Hill Tea Table. This is a bedrock-texture-remnant pillar.

This small pillar stands on the lower slopes of Fort Hill at the very rim of Baker Fork Gorge. This puts it near the top of the Peebles platform which is marked through most of the gorge by the tops of dolomite bluffs. Here the top of the Peebles Dolomite appears as a narrow, triangular flat area between two short, steep gullies which fall to the river below. Fort Hill Tea Table stands on this flat which extends on either side of it as a very narrow ledge. The pillar extends up into the so-called "marl" layer of the Peebles and represents a knot of harder dolomite which has been isolated by the erosion of weaker material from around it. Fossils of coral and apparent algal stromatolites — layered, rounded calcareous fossils formed by extinct colonial organisms — found in the table of the pillar indicate that it may represent a small reef-like structure called a biostrome. In spite of its vuggy nature, the rock forming the table has greater resistance to erosion than the leg and has protected the denser but weaker rock beneath it. The rounded outline of the tea table may be a result of the "teapot effect."

Both the table and the leg are elongated parallel to the neighboring gullies, and the leg thins noticeably toward the river. The present form of the pillar may be due in part to the action of these short waterways. On the other hand, the location of the gullies may have been determined by the presence of the resistant lump of rock forming the pillar. Perhaps they form a "geofeedback" system in which the gullies and the pillar affect and are affected by each other.

Pompeys Pillar (OH-P-GRE-01)* — Greene County
Height: 15 feet (4.6 meters); Width 3 feet (0.9 meter)
If Ohio has a famous natural pillar, this is it. Located on the grounds of Glen Helen, a private nature preserve owned by Antioch College in Yellow Springs, Pompeys Pillar (figures 167 and 168) has been a local attraction for well over a century. During the 1800s, Neff

Figure 167. Pompeys Pillar. This is perhaps Ohio's most famous pillar.

Figure 168. Pompeys Pillar viewed from the side (Van Tassel, 1901). This photograph, taken around 1900, shows that the pillar has changed little over the past century in spite of misguided efforts to topple it. The sloped bedding planes are a result of the pillar's slow migration downslope.

Park was established to take advantage of the water of the Yellow Spring for which the nearby village was named. Visitors came to the impressive Neff House hotel to "take the waters" and enjoy excursions through the woods and beside cliff-lined Yellow Springs and Cascade creeks. Along the way, they could see the Indian mound, listen to the music of The Cascade and view the impressive sight of Pompeys Pillar.

This tea table exhibits the same undercut profile as the Silurian dolomite cliff which stands behind it. Cedarville Dolomite forms a resistant cap 3.5 feet thick and 8 feet long protecting the less-resistant leg of Springfield Dolomite which has a width near the base of 8 feet, but narrows to a width of only 3 feet directly under the table. The impressiveness of the pillar is heightened by the thick platform of Euphemia Dolomite on which it sits.

The pillar most likely resulted from the enlargement of a vertical joint which isolated it from the cliff. There is, however, an unusual amount of space between it and its parent cliff. While erosion may be responsible for all of this space, a slight tilt noticed in the bedding planes of the pillar indicates that some of the distance may have resulted from downslope movement of the platform of the pillar. Such movement must have been exceedingly slow since the thin-layered Springfield Dolomite forming the leg of the tea table could not withstand much stress.

The formation of Pompeys Pillar is intimately associated with the formation of the gorge in which it is found. Water-worn gravel deposited by a meltwater stream coming off the Wisconsinan glacier has been traced from the head of the gorge upstream to a jumble of gravel hills located 2 miles northwest of the town of Yellow Springs. These hills, called "kames," were created when a meltwater stream running on the glacier's surface plunged into sinkholes melted through the ice. Much of the gravel it had collected from the top of the melting glacier was dropped at the base of the sinkholes. The water found its way beneath the ice to emerge at its front as a reconstituted stream with enough power to cut down into the dolomite bedrock and form the gorge of Yellow Springs Creek. When the melting front of the glacier retreated far enough, the piles of gravel at the base of the former sinkholes were revealed as kames.

Isolation of the pillar could have begun as the gorge of Yellow Springs Creek was being cut during the waning phases of Wisconsinan-stage ice retreat. However, much of the erosion which has left the pillar behind must have taken place after final retreat of the ice and through less violent agencies. This can, again, be deduced from the delicate nature of the Springfield Dolomite forming the leg. It would not long survive the active erosion of a raging stream filling the gorge, or the curiously destructive activities of college students, either. In 1973, shortly after Ralph Ramey became the director of Glen Helen, he was given a tour by the out-going director Ken Hunt. When they came to Pompeys Pillar, they found the college maintenance man pushing mortar into thin openings between the bedding planes under the massive cap rock, repairing damage caused by excitement-seeking visitors who pried rocks out in an attempt to get the cap to fall — a short-sighted, foolish and dangerous pastime. The mortar is still there, and so is the cap.

The name Pompeys Pillar has been used for this feature since the nineteenth century. No record has been found as to when or by whom the name was bestowed, nor is it known why it was chosen. One possible source is the account of the Lewis and Clark Expedition of 1804–1806 which was published in 1814. In it, William Clark tells of seeing a "remarkable rock" 200 feet high along the Yellowstone River. He called it Pompeys Tower in his journal, naming it for Sacajawea's infant son who he called Pomp. By the time his journals reached printed form, the name had become Pompeys Pillar. Given the popularity of the account of this adventure, it would not be surprising if a local "remarkable rock" were given the same name in emulation of the larger one.

John Bryans Window (OH-P-GRE-02) — Greene County
Height: 7 feet (2.1 meters); Width: 2.3 feet (0.7 meter)

A narrow fissure 4 inches wide separates the table of John Bryans Window (figures 169 and 170; Plate 16) from the adjacent cliff, making a pillar of what at first glance appears to be a small arch, a deception further encouraged by a separate slab of rock which spans the gap at its top. This pillar is a classic tea table having a wide top supported by a narrower leg. Located on the rim of the cliff overlooking the gorge of the Little Miami River in John Bryan State Park, it represents a form midway between Tecumseh Arch and Pompeys Pillar.

As with Tecumseh Arch and Pompeys Pillar, the massive Cedarville Dolomite, here narrowed to a thickness of 5 feet, provides a protective cap to a pedestal of thin-bedded Springfield Dolomite. The difference in resistance between the two layers is clearly displayed in the profile of the crevice separating the pillar from the cliff. In the Cedarville, this crevice is only 4 inches wide. In the Springfield below, it is 1.3 feet wide, forming the "window." Some of the

Figure 169. John Bryans Window. A view from the downslope side.

separation at the level of the Cedarville Dolomite may be due to failure of the Springfield Dolomite leg whose thin beds offer a number of planes along which slippage can occur. Added to this is its notably smaller width in comparison to the massive overhanging table. Gravity acting in conjunction with these two conditions has caused the pillar to tilt slightly downslope, opening the crevice separating it from the cliff. The wider opening in the more thinly bedded and broken Springfield Dolomite, on the other hand, is due mostly to erosive agencies.

One active element in the formation of John Bryans Window not usually found associated with pillars in Ohio is an intermittent stream which tumbles over the cliff beside it. The water actually falls into an extension of the widened vertical crevice which separates the pillar from the bedrock. The downslope wall of the crevice has fallen away beside the pillar so that the stream, which had been channeled toward the window by the crevice, now turns ninety degrees to fall as

Figure 170. John Bryans Window showing the enlarged vertical crevice that separates it from the cliff.

a shallow cascade beside the pillar and then over the abrupt edge of the Euphemia Dolomite shelf on which the pillar sits. Not all of the water goes over the falls, however. Some of it seeps down through crevices in the Springfield and Euphemia dolomites to emerge as small springs along the base of the ledge on which the pillar stands. When in flood, the stream flows across the floor of the window itself. Spray from the cascade keeps the leg of the tea table wet, promoting weathering. Such active erosion will rapidly weaken the foundations of this interesting feature.

Humans are also aiding the disintegration of this tea table. It has been noted that some visitors cannot resist the temptation to pry out pieces of the leg, probably from a misplaced desire to see the cap rock tumble down the steep slope below. This willingness to sacrifice a state scenic treasure for a few seconds of "thrill" is, unfortunately, all too common.

Teakettle Rock (OH-P-ADA-01) — Adams County

Height: 16.5 feet (5 meters); Width: 7 feet (2.1 meters)

The appropriately named Teakettle Rock (Figure 171) is found at the top of a cliff of Peebles Dolomite overlooking Ohio Brush Creek within the Edge of Appalachia Preserve. It has the form of a rounded tea table or, as the name implies, a teakettle, 10.5 feet tall sitting on a platform which is 6 feet high on the cliff side and 29 feet high on the downslope side. This difference is characteristic of those pillars which are detached cliff projections.

Teakettle Rock is one of several pillars found in the cliffs along the eastern side of Ohio Brush Creek near its junction with the Ohio River. In preglacial time, a divide between two tributaries of the old

Figure 171. Teakettle Rock. Pete Whan of The Nature Conservancy provides scale in this view of the pillar taken from below.

Teays River system crossed what is now the valley at nearly a right angle. The glacial advance which destroyed the Teays system dammed the northern tributary and ponded its drainage against the divide. The rising water eventually overtopped the divide and cut through it. The resulting gorge became a conduit for local drainage forming Ohio Brush Creek after the retreat of the ice.

The teakettle shape of this pillar is due to varying resistances of different beds within the dolomite. In this location the usually obscure beds of the Peebles Dolomite are more distinct and weathering has emphasized their strengths and weaknesses in relation to each other. Only after jumping across the narrow gap separating the tea table from the cliff in order to measure its top did preserve manager Pete Whan and I discover how weak some of these beds were. Its narrow waist is riddled with solution holes, several of which penetrate it completely. We both decided that one trip out was enough. The table, measured at such potential risk, turned out to be roughly square and 10 feet wide. At its closest approach to the cliff, it narrows the separating crevice to a mere 6 inches.

Wood Rat Tower (OH-P-ADA-02) — Adams County
Height: 23 feet (7 meters); Width: 10 feet (3 meters)
Wood Rat Tower (Figure 172), also located in the Edge of Appalachia Preserve, is one of the tallest pillars in Ohio. It is a massive remnant of the cliff, now separated from it by an enlarged crevice approximately 6 feet wide. It has a triangular cross-section and is aligned parallel to the cliff face. Its position on the steep slope which falls from the foot of the cliff gives its outer face an impressive rise of 46 feet. It was named by Pete Whan, preserve manager, for a possible nest of the endangered Eastern Wood Rat (*Neotoma floridana*) found in an alcove at its base.

Prow Tower (OH-P-ADA-03) — Adams County
Height: 19 feet (5.8 meters); Width: Approximately 3 feet (0.9 meter)
Like Wood Rat Tower, nearby Prow Tower (Figure 173) is also triangular in cross section, but more narrowly so. Although its base is 15 feet long, it is only 3 feet wide at its broadest point and so this

Figure 172. Wood Rat Tower. This view, taken from below, shows the massive nature of the pillar.

Figure 173. Prow Tower. The resemblance to a ship which inspired the name of this pillar is more evident when it is seen in person.

pillar resembles a wall more than a tower. Prow Tower, unlike Wood
Rat Tower, is elongated perpendicular to the rise behind it and has the
appearance of being a solution remnant rather than an isolated section of cliff. Its sharp downslope edge which resembles the prow of a
ship is 35 feet high.

Cute Tower (OH-P-ADA-04) — Adams County
Height: 8 feet (2.4 meters); Width: 2 feet (0.6 meter)
    Found as it is in company with the massive pillars of the Edge
of Appalachia Preserve, tiny Cute Tower (Figure 174) deserves its
name. With an upslope height of 8 feet and a downslope height of 11
feet, it is by far the smallest of the natural pillars so far located in the
preserve. It rests on a base 7 feet long and only 2 feet wide.
    Cute Tower sits near the top of a semicircular amphitheater
carved into the upper levels of the cliff overlooking Ohio Brush Creek.
It is quite rough and pitted, and is an obvious remnant of Peebles
Dolomite which caps the hills in this region. The upland surface behind the pillar shows extensive solution features such as pits and enlarged crevices. The dolomite appears to have been more susceptible
to this form of erosion here than at most other places in the preserve.
The alcove in which Cute Tower sits may be more a product of solution than mechanical erosion.

Pillars in Mississippian Strata

Big Pine Creek Pillar (OH-P-HOC-01)* — Hocking County
Height: 27 feet (8.2 meters); Width: 3 feet (0.9 meter)
    Big Pine Creek Pillar (Figure 175), located in Hocking State
Forest,.is certainly one of our more impressive natural pillars. This
tea table is especially notable for the height of its leg in comparison
to its width. It rises 27 feet on the uphill side and 32 feet on the downhill side from a base that is only 3 feet wide and 14 feet long. This
slender support narrows even more as it rises to a much wider table approximately 13 feet wide, 12 feet long and 5 feet thick. This results in an
impressive amount of overhang above a remarkably tall, thin leg.
    As with most other scenic features in the Hocking Hills region,
Big Pine Creek Pillar has formed in the Black Hand Sandstone. It is a

Figure 174. Cute Tower. The scale is 12 inches long.

remnant of the cliff 7 feet behind it, separated by a widened vertical fracture. The resistant layer of sandstone forming the table is evident in the top few feet of the cliff where it forms a slight projection. The lower, less-resistant layers forming the leg are undercut in the cliff. A short distance to the west, a very large cliff remnant displaying the same top-heavy profile, although in more massive form, has tipped back against the cliff to form an impressive gravity arch. Big Pine Creek Pillar was named for Big Pine Creek which flows in the valley below it.

Figure 175. Big Pine Creek Pillar. This is one of Ohio's more impressive pillars because of its size and the large overhang of the table compared to the narrow leg.

Hocking County Balanced Rock (OH-P-HOC-02)
— Hocking County
Height: 15 feet (4.6 meters); Width: 3 feet (0.9 meter)
The best, and perhaps only reasonably accessible, view of Hocking County Balanced Rock (Figure 176) is from the enlarged vertical crevice which leads to the narrow platform on which it stands. The rock itself is a vaguely spherical mass of stone 9 feet high standing on a blocky platform 6 feet tall. The connection between the rock and

302

Figure 176. Hocking County Balanced Rock. The pillar as seen from the enlarged vertical crevice. There is a 50-foot drop to its left. The overhanging edge of Flat Iron Rock is visible in the upper left hand corner of the photograph.

the pedestal narrows to 3 feet, making this pillar technically a tea table. The narrowness of this constriction is what gives the rock the appearance of being so delicately balanced that the slightest jolt would send it crashing to the slope below. It may, however, be sturdier than it appears from this angle. A clear view of the pillar's other sides is prevented by a vertical wall on one side and by a sheer 50-foot drop on the other.

The formation of this pillar is intimately associated with the enlarged crevice that defines its surroundings. The right wall of the access crevice extends to and beyond Balanced Rock, but the end of the left wall has been removed, leaving a small platform at its end on which stands the pillar. This cramped space not only holds Balanced Rock, but also Flat Iron Rock described below. The flat surfaces of the floor and extended wall of the access crevice, as well as the abrupt end of the left wall, indicate joint control. At first glance it would appear that a large block of the left-hand wall simply fell away. If that were the case, however, Balanced Rock and Flat Iron Rock would have been carried down with it. The missing rock must have been removed by atmospheric weathering which was halted or considerably slowed once it reached the joint faces forming the present walls on two sides of the platform. This conclusion is further strengthened by the heights of both pillars which equal that of the crevice wall beside them, indicating that they are actually remnants of it.

Balanced Rock is apparently an example of a bedrock-texture pillar — a more tightly cemented area of bedrock left behind by erosion of the less-resistant rock surrounding it. The constricted neck represents a layer of weaker rock between the large block above and the platform below.

The access crevice connects at its opposite end to a much longer vertical crevice barely 3 feet wide, indicating that the complex surroundings of the two pillars are actually located in one end of a massive sliver of Black Hand Sandstone separated from the main body of bedrock. M Arch is found in the other end of this monolith.

Flat Iron Rock (OH-P-HOC-03) — Hocking County
Height: 15 feet (4.6 meters); Width: 1.5 feet (0.5 meter)
Turn away from viewing Hocking County Balanced Rock and you find your nose pressed against the sheer face of Flat Iron Rock (figures 177 and 178). This unusual pillar consists of a flat, triangular slab of rock approximately 3 feet wide, 10 feet high and over 20 feet long balanced on a short neck above a platform 5 feet in height. The thin layer of rock forming the neck has eroded inward on all sides, creating Ohio's most unusual tea table. Although the neck is 16 feet long on the broad side of the rock, it is only 1.5 feet wide beneath its narrow side. The overhanging table extends a considerable distance beyond both sides of the constriction, causing one to wonder how this massive block maintains its stability.

But wait! There's more! When viewed from the narrow side of the pillar, the table's overhang is seen to be even greater. This is not due to the table becoming wider, but to the neck becoming even

Figure 177. Flat Iron Rock. This composite picture shows the entire broad face of Flat Iron Rock. Hocking County Balanced Rock is directly behind the photographer. The photograph in Figure 178 was taken while standing in the widened crevice that gives access to this narrow ledge, barely visible behind the sapling on the left, on which these two pillars stand.

narrower. From this vantage point, the end of the neck fronting onto the access crevice where the width measurement was made is seen to be a bulbous swelling of rock. Beyond it the neck is narrowed even further by what amounts to a long, low, shallow alcove. The opposite side of the neck is not accessible due to a precipitous drop-off, but it can be seen and that view reveals a similar low alcove running along the same line. This exaggerated constriction has formed in the same thin layer of less-resistant Black Hand Sandstone as has the narrowed neck of neighboring Hocking County Balanced Rock. The back walls of the alcoves must be separated by mere inches, meaning that the massive rock "flat iron" above is indeed carefully balanced on a knife's edge (Figure 178). It may be that the rock swelling at the crevice end of the neck is acting as a wedge stabilizing what would otherwise be a very precarious situation, much like the old college trick of balancing a hexagonal salt shaker on a single corner of its base in a bed of salt and then carefully blowing the salt away until only a few grains remain to hold the shaker in place. Blow away too much salt or have a mischievous comrade jar the table and the whole trick collapses. This platform with two delicately balanced rocks of impressive size is no place to be during an earthquake.

Flat Iron Rock was named for its resemblance to the bottom of antique irons designed to be heated on a stove. The name is hardly creative. There are a number of Flat Iron Rocks across the country, the most famous being the series of huge up-thrust slabs veneering the face of the Front Range of the Rocky Mountains above Boulder, Colorado. Even so, it seemed appropriate to give Ohio a bit of this tradition.

Figure 178. Flat Iron Rock. The narrow end showing the deep undercuts on each side and the bulbous knot of rock acting as a wedge.

Pillars in Pennsylvanian Strata

Chimney Rock (OH-P-PIK-01) — Pike County
Height: 23 feet (7 meters); Width: 15 feet (4.6 meters)

Chimney Rock (figures 179 and 180) is one of the few representatives in Ohio of a true large chimney — tall, narrow and comparatively straight-sided. In climbing the steep ridge toward the crest where it stands, one finds the rock looming overhead, seeming to grow taller as it is approached from below. Although Chimney Rock is itself 23 feet tall, the steep hill on which it stands and its sheer vertical rise make it appear much taller.

Chimney Rock is a remnant of the Sharon Formation which outcrops near the top of the hills in its vicinity. The Sharon here reaches a thickness of 70 feet and forms a line of cliffs encircling the ridges. Erosional widening of intersecting fractures near the edges of the exposed conglomerate has led to the formation of large cliff outliers and slump blocks.

Figure 179. Chimney Rock. In this view of the broad side of Chimney Rock, the crack dividing the pillar is visible.

Figure 180. Swarming Chimney Rock. How did they get up there? Little Chimney Rock is visible at the right side of the photograph. (Use courtesy of Ohio Department of Natural Resources, Division of Geological Survey)

The ridge on which Chimney Rock stands is a spur extending south from a larger hill. The chimney represents the southernmost remnant of the Sharon Formation which caps this spur and the hill beyond. Cliffs marking the edge of this conglomerate cap begin a short distance upslope from the pillar, lining both sides of the spur and continuing around the main body of the hill. The cliff point on the spur itself is broken up by the enlargement of fractures and forms a small and very interesting "rock city," within which is Little Chimney Rock, described below.

The shape of Chimney Rock obviously has been determined by intersecting vertical fractures, as is shown by its straight sides.

Another such fracture splits the chimney from top to bottom. This fracture reaches a width of 2 feet in places, very nearly making two pillars out of one. Due to the inadvisability of climbing this tall feature, the height of Chimney Rock was inferred from measurements of the cliff behind it. The platform of the pillar is elongated parallel to the main axis of the slope on which it stands.

Chimney Rock was a noted landmark at least as early as 1874 when Edward Orton, one of Ohio's premier early geologists, mentioned it in his report on Pike County for the state geological survey. It was also mentioned by Wilber Stout in his 1916 report on the geology of southern Ohio. Both reports used the term "Chimney Rocks," indicating the authors' awareness that Chimney Rock is not an isolated feature. By the early 1900s the rugged hills of southeastern Ohio had been for the most part cleared of trees, making the hilltops readily visible. The size and commanding position of Chimney Rock would have made it a prominent and easily recognized feature, and thus an attraction for visitors, many of whom were evidently less timorous than the author (Figure 180). Now that the hillsides are once more covered with fair-sized trees, Chimney Rock has retreated into relative obscurity.

Little Chimney Rock (OH-P-PIK-02) — Pike County
Height: 11 feet (3.3 meters); Width: 5 feet (1.5 meters)

The Sharon Formation in the immediate vicinity of Chimney Rock is made of two distinct layers separated by a resistant, iron-impregnated bed a few inches thick. Chimney Rock itself has formed in the lower, pebbly layer and is somewhat protected from weathering by remnants of the thin resistant bed capping it.

Upslope from Chimney Rock, this lower layer has been broken into a "rock city" by the widening of intersecting vertical fractures to form narrow, steep-sided passages which isolate several large blocks from the main body of conglomerate. The tops of these blocks are determined by the horizontal iron-bearing bed. Rising above it are remnants of the coarse sandstone upper layer of the Sharon Formation which have formed a "table," a short "mushroom" and Little Chimney Rock (Figure 181) on top of the isolated blocks.

Figure 181. Little Chimney Rock. Little Chimney Rock is on the right and the "mushroom" stands to its left.

Like Chimney Rock, Little Chimney Rock is a straight-sided pillar elongated in a down-slope direction. It is 5 feet wide and 9 feet long at its base and rises 11 feet above the top of the block on which it stands. Its original form was determined by vertical fractures, but its smaller size and more exposed position have allowed weathering to reduce it to a more cylindrical form than that attained by Chimney Rock.

Morel Tea Table (OH-P-PIK-03) — Pike County
Height: 20 feet (6.1 meters); Width: Approximately 6 feet (1.8 meters)

Morel Tea Table (Figure 182) is located north of, and on the same ridge as, Chimney Rock, and also has formed in the Sharon Formation. It is a massive cliff remnant rising from the steep slope at the base of the cliff with a height of 20 feet and an impressive downslope rise of about 40 feet.

The two-layered structure of the Sharon Formation found on this ridge is quite visible in the pillar. The light brown lower layer forms a slightly undercut leg. The upper layer swells outward due to a narrow, iron-rich seam of greater resistance which supports it. This bulbous table is darker brown and more deeply pitted than the leg, giving this pillar the appearance of a giant morel mushroom of the type popular with wild-food enthusiasts. The leg of the pillar is elongated parallel to the cliff behind it, giving it a length of 13 feet and a width perpendicular to the cliff of approximately 6 feet. The table extends several feet over the elongated ends of the leg, but is actually narrower than the leg on the front and back sides, making this pillar something of a hybrid between a tea table and a chimney.

The crevice separating Morel Tea Table from the cliff is 6 feet wide at the base. The tower leans slightly toward the cliff, apparently the result of erosion, not movement. The result is that in passing

Figure 182. Morel Tea Table. An enlarged vertical crevice separates this pillar from the cliff on the right.

through the crevice and looking up, one gets the uncomfortable impression that the whole thing could fall in at any minute.

Mineral Tea Table (OH-P-ATH-01) — Athens County
Height: 10.5 feet (3.2 meters); Width: Approximately 5 feet (1.5 meters)

Here is the classic tea table. A roughly circular table 13 feet in diameter and 3 feet thick is supported by a pedestal 8 feet tall. Formed in the Lower Freeport Sandstone, Mineral Tea Table (Figure 183) is a result of differential erosion. A resistant upper layer forms a wide, protective cap over the narrower leg formed in thin-bedded, less-resistant layers. It stands on a platform of massive sandstone slowly eroding out of the hillside which gives it a downslope rise of 15 feet. It is named for the nearby town of Mineral and is little more than a mile from Mineral Arch. This is one Ohio natural pillar that has gained a small measure of international fame; a picture of it has been used to illustrate the *Wikipedia* internet article on tea tables.

The pillar is located on a spur extending from a ridge above Hewett Fork and marks the end of a layer of massive sandstone which

Figure 183. Mineral Tea Table. This pillar exhibits the classic form of a tea table.

outcrops farther upslope. Although it resembles a short mushroom, this tea table's location at the top of a ridge and its wide overhang which exceeds 5 feet in places make it an impressive sight. The top of the table, defined by a bedding plane, is quite flat. According to local tradition, this makes it the perfect place for the devil to treat his horde of imps to a potent hemlock brew, which led to its being called Devils Tea Table. However, Ohio has several pillars named Devils Tea Table, and so that name has been reserved for the now-vanished pillar in Morgan County. At the time of my first visit, a crude ladder leaning against Mineral Tea Table provided access to the top. It must be a peaceful place to contemplate the larger questions of life when the imps are gone.

Dale Tea Table (OH-P-WAS-01) — Washington County
Height: Approximately 20 feet (6.1 meters); Width: Approximately 3 feet (0.9 meter)
Dale Tea Table (figures 184 and 185), named for a nearby cross-roads village, takes the prize for having the strangest shape of any natural pillar in Ohio. From one direction it looks like a duck sitting

Figure 184. Dale Tea Table. The "duck on a rock" view. (Photograph by Waldo Morn; use courtesy of Ohio Department of Natural Resources, Division of Geological Survey)

on a pedestal (Figure 184). From another angle it resembles a large-headed falcon sitting on a massive block of sandstone (Figure 185). The head of the bird is a balanced rock approximately 4 feet thick which extends beyond the narrow neck beneath it on all sides. This neck is made of thin-bedded sandstone which has weathered back more than the monolithic head.

A thick column of massive sandstone forms the body of the falcon, an illusion which its rounded form would perfectly achieve were it not for a strange angular "bustle" jutting out from one side. At the

Figure 185. Dale Tea Table. This is the "falcon" view, showing the "bustle."

base of the column, tall boulder-like projections form the bent legs of the bird. Viewed from a certain position, two of these are visible; there are, however, three of them, making this a most unusual bird indeed.

The entire figure, which is approximately 15 feet tall, rests on a platform of massive sandstone 5 feet thick, giving the entire pillar a height of 20 feet. It is carved from one of the higher sandstone units of the Monongahela Group, possibly the Mannington Sandstone.

Although Dale Tea Table is obviously a result of differential weathering in a unit of multi-layered sandstone, the details of its formation are somewhat obscure. Like Pompeys Pillar in Greene County, it sits on a talus slope beneath an escarpment, in this case the steep wall of the narrow valley of a small tributary of the West Branch of Wolf Creek. Several large slump blocks in the area show that mass wasting, the downslope movement of large masses of rock, has been a factor in widening the valley. The platform of the tea table resembles these slump blocks in form and size, and would be readily dismissed as such were it not for that immense rock bird perched on its back. Faintly etched bedding planes on its downslope face are slightly tilted which could be evidence of cross-bedding. The tilt could also indicate movement of the block; it may be that we are dealing with a slump block that just happens to be carrying a pillar with it during its slow journey downslope, much like what is happening with Pompeys Pillar in Greene County. If this is the case, the movement must be gentle indeed. The block "head" appears to be so delicately balanced on the narrow neck that any abrupt motion would tumble it from its perch. A near-vertical joint passing completely through the leg has been widened enough to allow light to be seen between two of the boulder-like "bird legs." This also weakens the fabric of the pillar and makes it more liable to destruction if stressed. It would appear that any movement which is occurring must be extremely slow and gentle. More likely, the pillar formed where it stands. Either way, the ever-narrowing neck and the continual widening of the vertical joint foretell the ultimate end of this most interesting feature.

The Devils Tea Table (OH-P-MRG-01) (destroyed)
— Morgan County
Height: Approximately 25 feet (7.6 meters); Width:
Approximately 5 feet (1.5 meters)

The Devils Tea Table (figures 186–190) must have been a spectacular sight, a gangly mushroom of rock standing in isolated splendor on its lonely hilltop 200 feet above railroad carriages and canal boats and dusty pedestrians following the valley of the wide Muskingum River below. We know that it attracted attention, for travelers reportedly stopped as a matter of course to observe it and it was a favorite picnic site for those who lived near enough to make the trip.

Judging by the efforts of writers and photographers, this was perhaps the most imposing pillar in Ohio. More descriptions and pictures of it have been found than for any other natural pillar or arch in the state. Henry Howe, Ohio's tireless early historian, described it thus:

> One of the most remarkable natural curiosities of the Muskingum Valley is the "Devil's Tea Table," which stands on one of the bluffs on the east side of the river . . . Its position is exactly central on the top of a high hill, the ground sloping rapidly from it in every direction. It stands like a lone sentinel, keeping its silent watch, as the years go by, over the beautiful river whose waters glide by it on their way to the ocean (Howe, 1896).

Although from this description it would appear that this tea table was perched at the very top of a hill, it actually stood on the crest of a lower spur. Howe included a lengthier description submitted by Dr. H. L. True of McConnelsville, which deserves to be quoted in full:

> It consists of an immense table of sandstone estimated to weigh over 300 tons, supported by a slender base of shelly slatestone. It maintains its place and position mainly by its equilibrium, the top being so evenly balanced on the pedestal that if a small portion were broken from one side of the table it would cause it to topple over. The table is quadrangular or diamond shaped, and has the following dimensions: it is about 25 feet high, 33 feet long, 20 feet wide, 10 feet thick, and 85 feet in circumference. The dimensions of the base are as follows: length, 18 feet,

width 5 feet, height about 14 feet, circumference 40 feet. The long diameter is in a direction north and south. When this massive stone is viewed in close proximity it appears to lean in every direction, so that on whatever side an observer may be, it seems liable to fall on him (Howe, 1896).

It is not known how Dr. True obtained these dimensions, but it is assumed that he or someone whose ability he trusted actually measured the pillar. A photograph published in 1901 gives the best view of this truly monumental natural feature (Figure 186). It illustrates the accuracy of the impression that the rock appeared ready to topple over when viewed from below. This massive erosion remnant was so impressive that another doctor by the name of James Ball Naylor was driven to poetry:

> O monster rock! Firm-poised it stands
> Upon a base of crumbling shale;
> Twas shaped by Satan's cunning hands
> In ages past — so runs the tale —

Figure 186. The Devils Tea Table (Van Tassel, 1901).

And served Hell's demons, great and small,
As table to their banquet hall.
Though countless years have rolled away,
The Devil's table stands today
As firm as when, with hellish glee,
The black imps held their revelry.

It seems the feeble, flut'tring breath
That issues from the lips of death —
The faint and fickle summer breeze
That stirs the blossoms on the trees,
Could shake the great rock's slender base
And hurl it from its resting place;
And yet the strongest gales that sweep
Across the torrid Indian deep,
The Polar winds — the fierce cyclone —
Are all too weak — combined, alone —
To cast the monarch from its throne.

Beyond the blue Muskingum's bed
It rears its gray and wrinkled head:
Though aged, still erect — sublime,
It gazes on the march of time,
And towers above the verdant sod,
A monument to nature's God
When years on years have hurried past
Until God's dial marks the last;
Oh! May the grim old rock still keep
Its Vigil on the stony steep (Naylor, 1960).

Others were not so sanguine about the pillar's longevity. As early as 1873, E. B. Andrews, writing in the *Report of the Geological Survey of Ohio,* Volume 1, describes The Devils Tea Table as:

> . . . the remnant, or outlier, of a sandstone stratum resting upon shales. The shales have been disintegrated and largely removed, as also has the lower and softer portion of the sandrock. This work of disintegration is now going on, and probably before

many years the narrow base of the pyramid will give way, and the huge rock will go thundering down the hill on one or the other side of the narrow ridge (Andrews, 1873b).

The Devils Tea Table appears to have been the exception to the rule that large arches and pillars will not form in shale. However, it cannot be denied that the shale base of this massive "inverted pyramid" as Andrews called it was a fatal weakness. Doctor True gives some very interesting information concerning efforts to exploit that weakness:

> There is a difference of opinion as to whether this rock can be made to vibrate or not. Some claim it is easy to vibrate it while standing on top. My own experience is that it cannot be made to vibrate with a pole from the ground, although it looks as if it could be done (Howe, 1896).

In addition to the questionable wisdom of trying to vibrate a 300-ton tea table while standing on top of it, other reports state that during picnics at the site younger members of the party danced cotillions on its top (Figure 187)! True's attempt to disrupt the apparently delicate balance of this large tea table illustrates a tendency common in the nineteenth century and unfortunately not unknown in this one. It is graphically illustrated by the following anecdote, also given by True:

> In 1820 a number of keel-boatmen, under the direction of Timothy Gates, gave out that on a certain day they were going to push it down into the river. Many of the early settlers gathered there to witness the proceeding, but the boatmen failed in their attempt to unsettle it, and the crowd was disappointed. Several attempts to overthrow it have since been made, notably one by falling a tree against it, but all resulted in failure (Howe, 1896).

Thank goodness! For the sake of a few seconds of questionable delight, some people would destroy the work of ages. It would be satisfying to report that the same forces which created The Devils Tea Table determined the time of its end when erosion finally removed enough of the leg to cause the pillar to become unbalanced. The truth is apparently more sinister; witless humans finally prevailed. Early in July, 1906, it was felled by a charge of dynamite in celebration of the nation's glorious independence.

Figure 187. The Devils Tea Table while standing. Although these nine-teenth-century visitors are not dancing, they did manage to get on top of the tea table (Howe, 1896).

Although The Devils Tea Table has fallen, it has not disappeared. A barren patch of crumbled, thin-bedded grey shale marks the spot on the ridge where it stood (Figure 188). Slightly uphill is a bedrock outcrop of the same sandstone that formed the table section of the pillar. The Devils Tea Table was a remnant of this outcrop, isolated when erosion widened vertical crevices.

Most interesting of all is the table itself which lies at the bottom of the ridge where it landed, apparently right-side-up, on that July day many years ago (figures 189 and 190). Although now it looks like one more of the many slump blocks littering the slope, anyone familiar with the old photographs taken before its fall would recognize it immedi-ately. Initials carved on its top and sides strengthen the identification;

Figure 188. The Devils Tea Table site after the feature was destroyed. The narrowness of the ridge on which this pillar stood above the Muskingum River is readily visible.

Figure 189. For a short time after its demise, The Devils Tea Table still attracted popular attention. This view is from a postcard in the author's collection.

Figure 190. A recent view of the fallen Devils Tea Table. Note the initials carved into the "table" when it was still an impressive local landmark.

visitors would have little reason to expend such effort on a slump block. By actual measure, the table is 27 feet long, 9 feet wide at the narrow end, 14 feet wide at the larger end and 7 feet thick. Some of its reported 10-foot thickness was apparently made up of thin layers of sandstone, some of which now lie tumbled against it. In comparing this massive block of sandstone to the narrow ridge on which it stood, it is easy to see why it would appear ready to topple over no matter which side one viewed it from. In order to pass it, a visitor would have to walk directly beneath it or make a strenuous detour farther down the steep slope of the ridge. Considering the weak base of shale on which it stood and the narrow pedestal which supported it, The Devils Tea Table must have remained upright through delicate balance and sheer good luck.

Photographs of the tea table show it standing in a cleared area, readily visible from both the road and the river below. Such an impressive feature silhouetted against the sky naturally attracted its own set of legends. Dancing seems to play an important role in several of them. For example, the devil was said to come out of a nearby cave — since collapsed, of course — at night to dance a jig on the tea table, changing himself into weird shapes and uttering strange noises as he did so — something of a Midwestern "Night On Bald Mountain." Indians supposedly built fires on it and had beautiful maidens in fantastic dress dance around it in order to attract braves passing on the river below who would then be ambushed by warriors hiding nearby. In more modern times, young men were told that if they climbed to the top of the table and walked three times around its edge backwards without feeling fear, they would succeed in whatever they attempted. One can only assume that any such success they may have enjoyed would have little to do with common sense. Ohio is certainly the poorer in having lost this impressive natural feature.

Pillars in Permian Strata

Dunbar Tea Table (OH-P-WAS-02) — Washington County
Height: 23 feet (7 meters); Width: 5 feet (1.5 meters)
Dunbar Tea Table (Figure 191) is another expression of erosion in rocks of differing resistance and is located, as are many of Ohio's natural pillars, on a spur extending out from a ridge. In this case, the spur has been formed by the junction of two minor streams. The table

Figure 191. Dunbar Tea Table. This view from below shows the blocky nature of this pillar.

of this pillar is 8 feet thick, 11 feet wide and 13 feet long. These dimensions make it appear to be quite heavy in relation to its overall height.

Fern Pillar (OH-P-WAS-02)* — Washington County
Height: 15 feet (4.6 meters); Width: 9 feet (2.7 meters)
Fern Pillar (Figure 192) is located within sight of Irish Run Natural Bridge. You see it and a neighboring bulky standing stone (Figure 232) down the slope to the left as you approach the arch. This pillar has a thinner profile than its dimensions would indicate, especially when viewed from the downslope side where it has a 29-foot vertical rise. Although made of the same Permian sandstone as the lintel of Irish Run Natural Bridge and the nearby standing stone, Fern Pillar exhibits little of the massiveness of those two features. It was named for a lush growth of ferns found on a ledge on its side.

Potential Additions to the Catalog of Ohio's Natural Pillars

Most of the following potential pillars and standing stones have not been visited during this survey to determine if they meet the

Figure 192. Fern Pillar. Irish Run Natural Bridge is located approximately 100 yards behind the point from which this picture was taken.

criteria for listing or, indeed, if they still exist. Some have been seen, but for various reasons could not be measured.

Cedar Fork Standing Stone — Adams County

The small Cedar Fork Standing Stone (Figure 193) was viewed and photographed by the author during an early visit to Cedar Fork Arch. It was, unfortunately, not measured at that time and could not be relocated during a more recent visit to the area. The man in the photograph is 6-feet-2-inches tall, showing that the rock meets the height criteria for listing as a natural pillar, but appears to be too wide in relation to that height to qualify.

Shadow Rock — Athens County

Parsley (1989) reports that Shadow Rock, located in Rome Township on the south side of the Hocking River, is 30 feet high which would make it taller than any reported natural pillar in Ohio except

Figure 193. Cedar Fork Standing Stone. Jeff Knoop with The Nature Conservancy provides scale.

for the 80-foot "rock" in Coshocton County. Parsley's figure most likely reflects the height of the downslope side of the feature. An article on Athens County history may be the original source of the estimate. It calls the feature "Shadow Rocks," indicating that there are actually two pillars at the site — the main rock and its "shadow." An accompanying picture does show one fairly broad pillar with a much narrower one behind it.

Chimney Rock Hollow — Athens County

The US Geological Survey's Mineral, Ohio, quadrangle shows "Chimney Rock Hollow" opening onto Hewett Fork along State Route 56 between the towns of Hocking and Mineral. Employees of the Waterloo Wildlife Experiment Station on whose grounds most of the hollow lies indicated that there was indeed a rock in the hollow which looked like a chimney.

Small Standing Stone — Belmont County

During an archaeological survey of the south-central portion of section 30, Warren Township, Murphy (2004) noted a small, 5- to 6-foot-tall pillar of Uniontown Sandstone on a ridge at the head of a small stream leading into Spencer Creek. Since the survey was performed as part of a strip mine permit application, this pillar most likely no longer exists.

Goblet Rock — Belmont County

In the *History of Belmont and Jefferson Counties, Ohio* by J. A. Caldwell (1880), a pillar called "Goblet Rock" (Figure 194), described as being 9 feet high and 31 feet in circumference around its flat top was said to stand on the Riggs farm 2 miles south of Barnesville. This is presumed to be the same feature called "Devil's Tea Table, South of Barnesville, Ohio" in an old postcard in the author's collection. Murphy (2004) was not able to locate the feature.

Rock — Coshocton County?

A reader responding to an article in the January 1988 issue of *Wonderful West Virginia* mentions a rock on the farm he grew up on near Coshocton, Ohio, that was 12 feet wide at the base, 1.5 feet wide at the top and 80 feet high. This sounds more like a skinny pyramid, and the height as given must be either a typographical error or a case

Figure 194. Goblet Rock, also called Devils Tea Table, from a postcard in the author's collection.

of expanded memory since it far exceeds the height of any known pillar in Ohio. One suspects that 8 feet would be closer to the truth. This may or may not be the same feature as the Baltic Pedestal Rock.

Gigantic Pillar — Geauga County

Warren Phillips Luther in a 1988 *Pholeos* article entitled "Some Interesting Ohio Caves in Noncarbonate Rocks" discusses several gorges cut into the Sharon Formation of Sand Hill near the town of Fowlers Mill. In one of them, which he calls Twin Falls, a "gigantic 'pillar'" divides a small creek into two waterfalls.

Bird Run Standing Rock — Guernsey County

An early (post 1910) postcard led James Murphy (2004) to Bird Run Standing Rock (Figure 195) on a narrow ridge between two tributaries of Wills Creek 1 mile north of the village of Birds Run,

Figure 195. Bird Run Standing Rock. (Photograph by Russell Pryor; use courtesy of Ohio Division of Geological Survey)

Wheeling Township. A recent photograph and that on the postcard show this pillar to be a true tea table, although one with a very short leg and a very thick, rugged table. Using a man standing beside it in the postcard picture for scale, the pillar appears to be about 15 feet high and 10 feet wide at the top. As with many places in this part of Ohio, strip mining has encroached upon its surroundings. Murphy found the area grown over and the rock difficult to photograph. The picture used here was submitted as part of an Ohio DGS contest and was taken when this potential pillar was easier to see.

Prider Road Pillars — Guernsey County

In northwestern Guernsey County stands an unusual cluster of pillars readily visible from Prider Road. Aligned on top of a narrow ridge, these four pillars were apparently carved from the Buffalo Sandstone of the lower Conemaugh Group according to Murphy (2004). There is nothing in Murphy's photographs to indicate height, but the stones appear to be 10 to 15 feet tall. One of them evidently has an arch eroded through it, giving the site added interest.

Cadiz Standing Stone — Harrison County

According to Joseph T. Harrison (1922), Cadiz Standing Stone (Figure 196), located 1 mile west of Cadiz in the northeast quarter of section 11, Cadiz Township, was the best known of three such rocks located in the county. It was composed of sandstone and would have been about 20 feet tall before "the desecrating hand of the white man" took off about 8 feet from the top for building purposes. Harrison estimated its height at the time of his visit at 12 feet. Its top was 16 feet wide and 18 feet long. Murphy (2004) noted that the stone stood just north of the old Standingstone Indian Trail and was probably a landmark for ancient travelers. He also found that the desecrating

Figure 196. Cadiz Standing Stone, as portrayed in a postcard in the author's collection.

hand of the white man had completed its work; the Cadiz Standing Stone has been destroyed by strip mining.

Scio Stone — Harrison County

Also reported by Harrison (1922), the Scio Stone is located 2 miles northeast of Scio in the northeast quarter of section 27, North Township. In 1922, it was readily visible on a hilltop to the left of the road between Scio and Kilgore near its junction with the road leading to New Rumley, proudly noted as the birthplace of General George Armstrong Custer. This pillar also had about 8 feet removed from its top and would have originally been 20 feet tall. A photograph of the Scio Stone accompanying the article shows the dapper, white-collared author standing alongside it. Harrison claims it to be 10 feet high and 18 feet by 12 feet on top. It was still in existence when Murphy visited it.

Indian Watch Tower — Harrison County

Harrison's third pillar is located in section 6, Moorefield Township, one-quarter mile north of "the old Nottingham Church" and 3 miles northeast of Moorefield. He estimated its height at 18 feet and the dimensions of its top as 10 feet by 10 feet. Unlike the other two pillars in Harrison County which are located on hilltops, the Indian Watch Tower stood about 50 feet below the top of the nearest hill. A photograph accompanying the article shows this pillar to be a true tea table with a very thick, blocky table and a short, constricted leg. When visited by James Murphy, strip mining had advanced to within a few feet of it.

Ironton Boulder — Lawrence County

The photograph of Ironton Boulder shown in Figure 197 was taken around 1900 and appeared in Van Tassel's *The Book of Ohio*. Viewed from this angle, it appears to be an attenuated tea table, but may actually extend a fair distance beyond the people at its base and so be too broad to qualify as a listed natural pillar. The only way to ascertain its status is to measure it. Unfortunately, the only information given is in the caption beneath the photograph which simply states, "Boulder seven miles north of Ironton, Lawrence County." This vague location lies deep in strip-mining territory, and the boulder might

Figure 197. A pleasant outing at the Ironton Boulder (Van Tassel, 1901).

have been blasted away long ago. Judging by the man standing at its base, the "boulder" appears to be approximately 16 feet tall, 5 feet wide at the base and 8 feet across its bulbous top.

Pedestal Rock — Lawrence County

A sandstone "pedestal rock" noted in the files of Ohio DNAP is located in Rome Township, Lawrence County, near the junction of County roads 73 and 72. It is reported to have a table approximately 15 feet by 20 feet and 4 feet thick on a leg which is 5 feet tall and 5 feet thick.

Noble County Rocks — Noble County

All that is known of the two standing Noble County Rocks (Figure 198) is that they are, or were, located on Cline's Farm in the early 1900s. The photograph is a copy of an old postcard given to James Murphy.

331

Figure 198. Noble County Rocks. (Use courtesy of James L. Murphy)

Kent Standing Stone — Portage County

One of the most storied of Ohio's potential natural pillars is the Kent Standing Stone (Figure 199) located in a gorge cut by the Cuyahoga River at Kent. It was an important landmark along the old Mahoning Trail and tradition has it that Samuel Brady, a frontier scout, made a desperate leap to its summit and then to the far bank of the river in an attempt to escape pursuing Indians. They, however, simply crossed at the ford lower down and Brady was forced into a nearby pond which is still called by his name. There he hid beneath a raft of driftwood to escape them. The gorge was widened during construction of the Ohio and Erie Canal, but the standing stone is still there.

Scioto County Devils Tea Table — Scioto County

A photograph (Figure 200) posted on the Portsmouth, Ohio, Public Library web site shows a Devils Tea Table which the accompanying caption says is located in Madison Township one-half mile south of Pike County and one-half mile west of Jackson County. The date of December 5, 1909, is hand-written on the picture. The formation appears to actually be two tea tables, one on top of the other. The top formation is estimated to be 10 feet high and the bottom one 4

Figure 199. Kent Standing Stone located in the middle of the Cuyahoga River.

feet, although it is unclear whether or not the entire height of the bottom section is shown in the photograph. What is shown is a crowd of 30 men, women and children standing on the bottom section of the tea table and 16 men and boys perched above their heads on the very top of the rock. Even without the people, this geological oddity is one of the most striking specimens of its kind in the state. A more recent, and less peopled, photograph of this feature (Figure 201) was provided by James Murphy.

Baltic Pedestal Rock — Tuscarawas County

Reported to the author by the photographer Ian Adams, this pillar is located off State Route 93 east of the town of Baltic.

Balancing Rock — Washington County

A report in the files of Ohio DNAP notes a "balancing rock" approximately 10 feet by 10 feet on a pedestal 5 feet in diameter in section 31, Wesley Township, near the junction of State routes 676 and 555.

Figure 200. Scioto County Devils Tea Table. An early twentieth-century version of "How many people can we get on a rock?" (Use courtesy of Portsmouth Public Library, donated by Debbie Keeton)

Figure 201. A recent view of the Scioto County Devils Tea Table. (Use courtesy of James L. Murphy)

Chapter 6

Fallen Arches
and Toppled Tables

All good things must come to an end and that includes geologic features. In the normal course of weathering and erosion they are whittled into rubble. Usually the process occurs at such a leisurely pace that it is not comprehended by human observers. However, under certain conditions geologic change may happen quite rapidly, even catastrophically.

Natural arches and pillars are prime candidates for catastrophic change. As erosion widens the opening of an arch or narrows the base of a pillar, it pushes the feature ever closer to the point of instability. Eventually a crevice is widened too far, a lintel is narrowed too much, a pillar becomes too lopsided and then gravity does the rest.

When a collapsing arch or pillar strikes the ground, the force of the impact usually shatters the falling rock into a pile of rubble. With more surface area exposed to weathering, the fallen pieces quickly disintegrate, leaving little to indicate that they were ever anything more interesting than a larger version of the rock pile they have become. However, in those cases where the collapse has occurred so recently that weathering has not yet obscured all evidence of the event, an opportunity is provided to study the last chapter in the history of these intriguing features.

The best such example in Ohio is found at Ladd Natural Bridge in Washington County. Just below the west abutment of the arch is a ledge paralleling the lintel above and equal to it in width (Figure 202). This ledge and the top of the natural bridge form the treads of a

Figure 202. A sketch showing the lintel blocks and abutment remnant (boldly outlined) of a possible fallen arch below Ladd Natural Bridge.

massive two-step staircase. The lower step, however, is incomplete, for the ledge ends in mid air. It is quite thick where it meets the abutment, but rapidly thins toward its hanging end, with the loss of rock occurring entirely on the underside. Seen from the valley below, the ledge appears to be an incomplete image of the arch above it. Could this be all that remains of a former natural bridge which may once have rivaled Ladd itself in size?

At the base of the cliff are two angular slump blocks comparable in size to the ledge remnant above. They lie in line with each other and with the ledge, and appear to be fallen pieces of a massive lintel. On the opposite side of the tiny valley formed by the intermittent stream which falls through Ladd Natural Bridge are several large, rounded blocks of stone which continue the line formed by the fallen blocks. There are more rocks in the center of the valley, but these are much smaller and more weathered than the others. This, however, is just what we would expect if this entire collection of slump blocks represents the remains of a fallen natural arch. These center blocks might have been thinner to start with. They certainly had a larger distance to fall and so the force of the impact would have been greater. Also, their position in the bed of the tiny stream falling from the ledge above subjects them to more rapid weathering.

So here we have evidence that Ladd Natural Bridge was, in the not-too-distant geologic past, a double bridge. Its fallen twin was made from the thick, resistant layer of sandstone whose top forms the floor of the existing arch. The cliff face of this lower layer has not eroded back very far, and so the skylight of the lower bridge could never have been more than a narrow fissure. It was evidently wide enough, however, to free the span and cause it to depend for support on the abutments. The flat faces of the most intact blocks of the possible span indicate that it was crossed in several places by joints which could have caused the bridge to fail if widened enough.

Although the existing natural bridge appears to be quite sturdy, it, too, will fail at some future date. However, groundwater seepage staining the face of the cliff below it shows that the processes which formed both the past and present arches still continue. This is one of the few sites found in Ohio where all three phases of arch history — incipient, present and collapsed — are present.

There are some other sites in Ohio that might represent collapsed arches. A bowl-shaped notch in the cliff edge next to Raven Rock Arch might be the remains of a similar feature. In the narrow valley below Three Hole Arch in Hocking County stands The Gateway (Figure 203), two points of rock which may be the remains of the abutments of a former arch, although the evidence here is far from conclusive.

While arch failure can result from the enlargement of joints and other lines of weakness, it can also occur as a result of void enlargement, either through the mechanics of a tension dome or through weathering. Should the top of an enlarging void reach the level of an arch's deck, the arch itself is in danger of collapse. At its simplest, arch failure can result from the slow process of weathering which whittles away at the expanse of unsupported rock until a breach is

Figure 203. The Gateway, possibly a collapsed natural arch, downstream from Three Hole Arch. The two vertical rock faces may be the remains of the abutments of an arch.

made. Almost Arch, located in Baker Fork Gorge, illustrates this form of arch failure (Figure 204).

Evidence of toppled pillars is even more difficult to find than is that for fallen arches. This is especially true for tea tables. When one collapses, its leg, which is often of thin-bedded rock, quickly breaks down while the more substantial table becomes just another slump block or shatters on impact. Fortunately, the larger pillars are often impressive enough to attract the attention of writers, artists and photographers who record their location and appearance. On-site comparison of these historic records with the present view can document changes that have occurred over a known length of time.

Some of the more dramatic examples of such recorded changes are to be found in the American West where pillars are both larger and more numerous. One of the most famous natural pillars in the country, Chimney Rock along the old Oregon Trail in Nebraska, has been drawn and photographed for well over 100 years (Figure 158). Comparing sketches of it made in the nineteenth century to later photographs, it appears that at some point, part of the top collapsed,

Figure 204. Almost Arch, located 10 feet downstream from Hidden Arch in Baker Fork Gorge, illustrates arch failure by means of weathering reducing the lintel until a gap occurs.

leaving a shorter pillar than that found by earlier travelers. A more recent example is found in Arches National Park where Chip-Off-The-Old-Block, a smaller version of the much-photographed Balanced Rock, fell during the winter of 1975–1976. Pedestal Rock in Indiana's Shades State Park is a nearer example. This short but massive tea table collapsed in 1975, leaving behind a pile of shattered sandstone slabs.

Ohio has its own collection of toppled pillars. The Devils Tea Table in Morgan County has already been described, but it is not the only example. Table Rock (Figure 205), a tea table located on the northernmost point of Kelleys Island in Lake Erie, was for many years a noted landmark. Pictured as early as 1840, this small but elegant feature was formed by the action of waves on the rocky foundation of the island. The Erie Islands and Marblehead Peninsula are formed of resistant Columbus Limestone and Bass Island Dolomite. Both beds dip downward to the east, quickly dropping below lake level. Thus while the west side of the islands and peninsula exhibit wave-cut cliffs,

Figure 205. Table Rock on Kelleys Island in Lake Erie (from a painting in the author's collection).

the east side slopes gently down into the lake. On the rocky western shores, it is common to find a thick, resistant layer of rock resting on thin-bedded, less-resistant rock which is often at wave level. The rock is further weakened by vertical crevices which often run at right angles to each other. The mechanical pounding of the waves coupled with constant moisture can rapidly erode rock exposed to such conditions. Frost action, probing roots and sheet ice shoved against the shore by wind and waves also do their share of destruction.

These conditions result in a shoreline which is constantly forming and then losing interesting geological features such as Table Rock. A set of joints crossing at nearly right angles was enlarged to define the table, and the thin-bedded layers beneath it were cut away by the waves to form the leg. The Erie shore is an energetic environment, however, and the same waves which crafted Table Rock eventually destroyed it. According to local informants, Table Rock fell sometime in the 1970s. The wonder is that it lasted so long, located as it was at one of the most exposed points of the island. The waves continue their work farther down the point, however, and new "table rocks" are being formed (Figure 206).

The complete story of another lakeside pillar has been captured on film. One photograph (Figure 207), taken in 1942, shows a prominent dolomite block jutting into Lake Erie from the cliffs near Catawba Point on Marblehead Peninsula, Ottawa County. Although joined to the mainland, the block's connection is being weakened by the erosion of a vertical joint which has already been widened to 3 or 4 feet. An aerial photo taken 26 years later (Figure 208) shows the elliptical top of the block completely separated from the mainland. The block has now become Flowerpot Rock (Figure 209), a pillar rising approximately 20 feet and worthy of some local notoriety. But even as the pillar attains its most dramatic shape, signs of its ultimate collapse are obvious. The ground-level photograph shows that the thinly bedded dolomite at the base of the rock has already been severely undercut by wave action. At the center, erosion has completely penetrated the base to form a small arch, creating an unusual two-legged pillar.

Figure 206. New table rocks being formed on the shore of Kelleys Island. Atmospheric erosion and wave action are widening intersecting crevices in the resistant upper level of rock and whittling away the thinly bedded rock supporting it.

Figure 207. A photograph of the western shore of Marblehead Peninsula taken in 1942 shows a spur of dolomite projecting into the lake. This feature later became Flowerpot Rock. (Use courtesy of Ohio Department of Natural Resources, Division of Geological Survey)

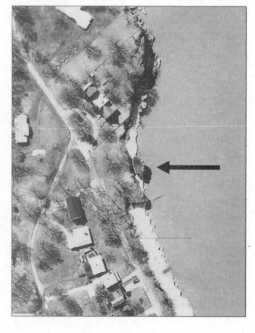

Figure 208. An aerial photograph taken in 1968 shows the spur of dolomite appearing in Figure 207 separated from the cliff. At this time it became known as Flowerpot Rock. (Use courtesy of Ohio Department of Natural Resources, Division of Geological Survey)

Figure 209. Flowerpot Rock as seen from the lake shore. Notice the small opening which has eroded through the thin beds forming the base of the pillar. (Use courtesy of Ohio Department of Natural Resources, Division of Geological Survey)

Figure 210. In this aerial photograph taken in 1973, Flowerpot Rock is gone. The arrow points to the remains of the collapsed pillar. (Use courtesy of Ohio Department of Natural Resources, Division of Geological Survey)

By 1973, Flowerpot Rock was gone; an aerial photograph taken in that year (Figure 210) shows only a pile of rock rubble where the pillar once stood. At some point in the intervening 7 years, the base of the rock failed and it toppled. Lake Erie quickly broke it to pieces.

Not all collapsed pillars have vanished so completely. One example (figures 211 and 212) is found in the bluffs of Peebles Dolomite fronting the Ohio River in Adams County. The rock here has been broken into a series of projecting headlands separated by erosion channels of various widths, as is typical of this rock. The headlands have been noticeably undercut due to a resistant cap layer above a less-resistant lower layer. This situation is not uncommon in the Peebles, but here the difference in resistance is greater or the erosive forces were stronger, for the overhangs are more extended than is usually seen. Where portions of the rock have been isolated and erosion of the lower layer can occur on all sides, the overhangs are noticeably wider on the side facing the river, possibly because of intensive erosion when the river was running at a higher level than it is now. On one of these isolated bluffs, erosion proceeded far enough on all sides to create a classic tea table. A narrow leg 6.6 feet wide and 5 feet long supported a 3-foot-thick table 9 feet wide and 21.5 feet long. From above, this table had the appearance of an elongated teardrop widest at the leg and tapering to a point toward the river. In keeping with the typical form of the region, the tapering point was longer than the overhang on the opposite side of the leg — 9 feet versus 6 feet. At some point this difference became great enough to destabilize the tea table and it fell over, separating itself from the bedrock at the base of the leg. It did not fall far, however. The formation was so short (only 6.6 feet high) and the extended point of the table so long, that the point hit the ground before the fall could generate enough force to shatter the formation. And so it leans, perfectly formed, but severed from its foundation, a true "toppled tea table."

Whether or not human activity played a part in toppling this tea table is not known, but would not be surprising. Destroying natural features for fun or profit has a long history in this country. We tend to think of geology as being permanent, at least in human terms, but such is not the case. Glacial kames and eskers are mined for the gravel

Figure 211. A toppled tea table in Adams County.

Figure 212. The base of the Adams County toppled tea table showing how it separated along a bedding plane.

they contain, cliffs are quarried away, rivers are dammed and straightened. Geology, too, must be made "useful."

Fortunately, some of the most spectacular geology finds its greatest usefulness in that very grandeur, attracting tourists and researchers to parks and nature preserves across the country. What would Yosemite and Yellowstone be without their geology? Natural arches and pillars possess a good deal of this attractiveness in their own right, in part because they are often picturesque and in part because they appear to be so unusual when compared to most of our experiences with geology.

These qualities are not always enough to protect them, however. The story of Mansfield Natural Bridge in Indiana should serve as a warning. This fair-sized arch crossed a small valley leading into Raccoon Creek, a stream that had the inconsiderate habit of rising over its banks every now and then. In order to tame it, the US Army Corps of Engineers proposed a dam. The best possible site was just downstream from the natural bridge, one of the few rock outcrops in the area capable of anchoring a dam. The landowner, hoping to take advantage of a seller's market in Indiana stone for building facings, opened a quarry to get as much profit as possible out of his property before the government acquired it and drowned everything. In the process, the natural bridge was found to be in the way and was quarried out.

As it happened, this particular stone was not very good for building purposes. Its color was drab and inconsistent and it spalled badly. The quarry closed down soon after it opened. The local rock was not suitable for anchoring the dam, either, and a site 2 miles upstream from where the natural bridge once stood was chosen. And so now instead of a scenic geologic feature of regional importance, there is only an overgrown, abandoned quarry pit. It is estimated that the stone from the natural bridge may have brought its short-sighted owner all of about twenty-five dollars.

Lest we gloat over the folly of our neighbors, let us remember David Road Natural Bridge near Dayton which was apparently destroyed to allow construction of a wider road, The Devils Tea Table on the Muskingum River blasted down as part of a Fourth of July celebration and the Cadiz Standing Stone, partly quarried away for building material and finally destroyed completely by strip mining. Pike Arch and Hole-in-the-Wall Arch are both endangered by the possible resumption of mining operations. Nor is outright destruction the only form of degradation such features can suffer. Bass Lake Natural Bridge is in danger of being buried by quarrying operations (Figure 213). Several of our more accessible arches and pillars have received the indignity of carved or spray-painted messages left by witless vandals, and litter is always a problem. The landscape we inhabit has preceded us by centuries and will outlast our transient lives by centuries more. It behooves us all to ensure that it does so in good condition.

Figure 213. Bass Lake Natural Bridge. A view of the arch against the high berm hiding a quarry which has encroached on the pit behind the natural bridge.

Chapter 7

Publicly Accessible
Natural Arches and Pillars
of Ohio

In Ohio, most land is privately owned. This means that the survival of the greatest part of our natural inheritance, including unusual or spectacular geology, depends on the wisdom and foresight of ordinary citizens. It also means that most of that inheritance is not readily accessible to those who do not happen to be the owner. For those interested in visiting Ohio's natural arches and pillars, however, the situation is a bit brighter. Well over a third of the features listed in this book are located by accident or design on publicly owned land, on land owned by private agencies that allow public access or near roadways from which they can be observed. In some cases, prior arrangements — such as the acquisition of a free access permit — must be made before the visit, but complying with these restrictions is rarely onerous. The extra measure of protection these regulations give these important features is well worth the modest extra effort required to visit them.

The following list gives information on those natural arches and pillars which are readily accessible to the public, enabling the visitor to become acquainted with some of Ohio's more notable examples of unusual geology. Only those publicly owned features which can be accessed without crossing private land and which have adequate, safe parking have been included. For convenience, the list is arranged by the agencies responsible for their protection. However, be aware that conditions change. Trails are moved or abandoned, agency regulations are revised, ownership of land changes. The information given here was accurate at the time of writing, but is subject to change, so visitors should contact the agencies responsible for each feature before visiting to obtain the most current information. Steep trails, stairways and other potential hazards are often encountered on these sites, and all due precautions should be taken. Honor all posted advisories and regulations to ensure a safe and pleasant visit.

Ohio Division of Natural Areas and Preserves
2045 Morse Road, Building F-1
Columbus, OH 43229-6693
614-265-6453
http://www.dnr.state.oh.us/dnap/

Features protected: Rockbridge, Saltpetre Cave Arch, Surprise Arch, Miller Natural Bridge, Miller Arch, Raven Rock Arch, Rockgrin Arch, Slide Arch, Ladd Natural Bridge and Cedar Fork Arch.

Ohio DNAP, a unit of Ohio DNR, is responsible for protecting outstanding examples of Ohio's natural heritage as outlined in the Natural Areas Act of 1970. It maintains a system of state nature preserves which includes old-growth forests, prairie remnants, wetlands, unusual wildlife habitat, intact ecosystems and geological features. These areas are not parks; rather, they are sanctuaries for rare plants and animals and are suited only for activities having minimal impact such as bird watching, hiking, nature study and photography. Ohio DNAP is also in charge of the heritage database which collects information on rare and endangered plants and other natural features. Most state nature preserves are officially dedicated; this is a legal designation that affords the highest level of land protection the state offers. The division does not own all the areas it protects. Land owned by other agencies and private citizens can be dedicated as state nature preserves with the consent of its owners if it meets the criteria for listing. In such cases, the scope of public access is determined by the property owner. For most of these, as well as for many division-owned preserves which are not able to handle unrestricted public visitation, an access permit is required. These permits are free and can be obtained by contacting Ohio DNAP. An application form is available on the DNAP web site. Directions to preserves requiring an access permit as well as any special instructions are mailed with the permit. It should be noted that pets and bicycles are not permitted in any state nature preserve. A photographic essay on Ohio's natural arches can be found on the division's web site.

The most easily accessed natural arch managed by the division is Rockbridge (Figure 86), which also happens to be Ohio's longest natural bridge. Open to public visitation from sunrise to sunset every day, this nature preserve is located off US Route 33, 1.5 miles southeast of the village of Rockbridge. Signs for Rockbridge State Nature Preserve direct the visitor onto Goodhope Township Road 124 and Falls Township Road 503, both announced by a large sign at their intersection with US Route 33. The road immediately turns east to parallel the federal route for about 100 yards, then turns north where a small sign shows that you have entered Falls Township and the road name has officially changed to Township Road 503 (Dalton Road). This road is very narrow and a little hilly, so drive with care. It ends within half a mile. The preserve parking lot is on the left.

The access trail from the parking lot to the preserve travels through what is essentially a 40-foot-wide fencerow. The worst of the wet places are crossed on boardwalks, but there are still plenty of opportunities to get muddy shoes. This protected "edge" where several different environments meet is very attractive to birds as shown by the tangle of blackberry, Japanese honeysuckle and multiflora rose, all of which are there because of the birds. The many habitats found in this preserve are one reason it has been made part of the Hocking Valley Birding Trail. Eventually the access lane widens into the main body of the preserve. The trail climbs the old valley wall of the now-buried preglacial ancestor of the Hocking River. At the top, you can look back over a fine view of the old valley which was much deeper before the glaciers filled it with rocky debris (Figure 214). Ahead, the trail drops down into the much narrower new valley cut by floods of meltwater coming from the ice sheet as it retreated.

The trail forks shortly after crossing the divide between the two valleys, but both branches lead to the same place — the natural bridge

Figure 214. A part of the valley of the preglacial Hocking River, now buried beneath rocky glacial debris. The hill on the right is the preglacial valley wall.

— and so make a loop. The left branch makes its way along the side of the hill, eventually dropping down into the shallow valley of a small stream tumbling toward the Hocking River. Do not be fooled by the small size of this stream; it is the architect of the impressive geological gem that you are about to see. Rockbridge sneaks up on you when approached from this direction. You will see it from above, noticing first that the stream appears to disappear over a ledge, and then realizing that it is actually dropping through a crevice several feet back from the ledge. This is the skylight of the natural bridge.

For those who prefer a more dramatic climax to their hike, taking the right-hand branch at the trail junction will lead through a young beech-maple forest along a ridge-top with good views down into the Hocking River Valley. At the end of the ridge, the trail descends a steep slope and here you catch your first glimpse of Rockbridge in all its glory. Both branches of the trail meet at the south end of the natural bridge. A short, steep trail leads farther down for a view of the underside of the bridge (Figure 215). This is as far as you can go;

Figure 215. Rockbridge as seen from the trail below it.

please do not cross the barricades that have been erected beneath the natural bridge and along the trail leading to it. They were placed there to protect the impressive wildflower display and scenic qualities of the area. There are plenty of good photo opportunities from the legal trails. Either branch of the trail will get you back to the junction. Which one you choose will depend on whether you prefer a gradual climb to the summit or the short, steep route.

Also found in the Hocking Hills region, Saltpetre Cave State Nature Preserve protects two arches. Location information and parking instructions are provided when the access permit is issued. Once there, follow the trail which eventually forks at a rock outcrop. The left-hand trail follows the base of the outcrop to Saltpetre Cave Arch which is a pillared-alcove-type arch (Figure 216). The right-hand trail follows the top of the outcrop and eventually drops down to Surprise Arch (Figure 110). This arch will be difficult to see. It is found in the wall between two large adjoining alcoves. Walk to the eastern alcove and then look back at the dividing wall. With luck, you will see a long, narrow opening. This is Surprise Arch.

The division also manages Miller Nature Sanctuary, a component of Rocky Fork Gorge State Nature Preserve where two arches are readily visible. This area requires a free access permit. Directions to the preserve and any special instructions will be mailed to you with your permit. Once there, leave your car in the parking lot and walk the access road which parallels Rocky Fork. This preserve is noted for its wildflower displays. In mid-March keep your eyes open for the diminutive Snow Trillium (*Trillium nivale*). Later in the year Shooting Star (*Dodecatheon meadia*), Wild Columbine (*Aquilegia canadensis*) and American Columbo (*Frasera carolinensis*) will be blooming. About half a mile in, you will pass a graveled parking lot used for special events on the left side. A short distance farther is a bulletin board with posted information and brochures on the sanctuary.

Figure 216. One of the two entrances to Saltpetre Cave Arch.

354

Figure 217. Miller Natural Bridge from the trail.

The skylight of Miller Natural Bridge is immediately to the right of the bulletin board. A short distance beyond the bulletin board, a sign indicating the junction of two trails marks the beginning of the trail system.

Arch Trail, one-third mile long, turns right down a small side valley into the gorge of Rocky Fork. Within a short distance the main opening of Miller Natural Bridge can be seen in the dolomite cliffs to the right (Figure 217). An interpretive plaque at the wooden bridge over the ephemeral stream coming down through the arch explains its formation. The trail continues along the bank of Rocky Fork for several hundred yards before coming to a junction which is the start of a short loop that passes around a large slump block called Flower Rock by the Miller family, previous owners of the property. In spring, the wildflower displays on the hillside between the trail and the base of the cliff are truly spectacular.

Back at the main trail junction near the bulletin board, the 1-mile-long Falls Trail strikes off to the left, crossing a sprightly (at

355

least in the spring) small stream on a bridge. On the far side of the bridge, the 0.75-mile-long Tuliptree Trail cuts off to head up the little stream and circle through a maturing stand of Tuliptree (*Liriodendron tulipifera*). Continue on Falls Trail which climbs out of the small side gorge onto the flats above where it passes through a middle-aged oak forest. At its farthest point, the trail abruptly drops into another side gorge, descending a long wooden staircase, and here Miller Arch makes its dramatic appearance (Figure 48). From the steps you get an impressive view of Rocky Fork below, framed through the arch. Another interpretive sign describes this arch. The wording on the two signs explaining the Miller arches may sound familiar; I did, after all, work for the division for a number of years. At the bottom of the steps the trail turns back upstream, passing through a rich bottomland community with Scouring-rush (*Equisetum hyemale*) and Virginia Bluebell (*Mertensia virginica*), and skirting several large slump blocks. A small side stream dropping over the lip of a shallow alcove forms the ephemeral waterfall for which the trail was named. Falls Trail eventually turns up the small stream valley it originally crossed and ends where it began at the junction with Arch Trail.

Raven Rock State Nature Preserve (access permit required) protects several natural arches which are accessible, but not easily so, for they are located at the top of a very steep and strenuous climb up a 500-foot-high bluff overlooking the Ohio River west of Portsmouth, Ohio (Figure 218). This hike should be attempted only by those in good physical condition since long stretches of the trail ascend at angles steeper than forty-five degrees. Be sure to take plenty of water.

The ascent begins almost immediately upon leaving the small parking lot, but do not be deceived by its gentle nature; it will get much worse. After passing through a tangle of young trees and multiflora rose, the trail enters a middle-aged oak woodland. Here the climbing becomes noticeably more strenuous, but it gets even worse ahead. On the steepest stretches, you can almost reach straight out and touch the trail ahead of you. Since you cannot change it, enjoy it. Take your time, rest often and concentrate on seeing the small treasures along the way — clusters of bluets, moss patterns, interesting stones, your feet. A rock outcrop on the left side of the trail marks the lower cliff

Figure 218. Raven Rock Ridge from below. Raven Rock and the arches are located on the highest knob.

line which can be seen stretching away along the face of the bluff facing the river. At the top of the outcrop, the trail levels out, following a narrow ridge that drops off to a steep slope on the right and to the lower cliff line on the left. Catch your breath here along this stretch, for at the end of the ridge the trail steepens once again. When you can see open sky and the framing of an interpretive sign ahead, you know you are nearly at the top. Once there, the trail breaks out of the woods onto the open platform of Raven Rock with a truly amazing view of the valley of the Ohio River. In spite of the hard climb, this view has attracted visitors for well over a century (Figure 219).

The interpretive sign gives a brief history of the area. The small rock jutting out into thin air just beyond it is Little Raven Rock. Raven Rock proper is the much wider flat rock to its right. Its cantilevered form is better seen from Raven Rock Arch located a few feet to the west (Figure 220). A devastating ice storm that struck southern Ohio in 2003, snapping off limbs and breaking down trees, was especially hard on this exposed bluff face. Once the tree canopy was torn open, brambles and shrubs — most notably poison ivy — were able to grow unchecked, making Raven Rock Arch less visible than it was before.

Figure 219. A less-than-intrepid nineteenth-century visitor to Raven Rock (Van Tassel, 1901).

Figure 220. Raven Rock and the Ohio River Valley. A view from the top of Raven Rock Arch.

The bowl of Raven Rock Arch's possible fallen twin is immediately to the west; the recently formed skylight of the emerging arch can be seen a few feet behind it.

Rockgrin and Slide arches (figures 120 and 121) are located below Raven Rock Arch and can best be reached by carefully descending through the bowl of the fallen arch to a sloping shelf found below. Follow this shelf as it gradually descends along the base of the cliff beneath Raven Rock. This is a good opportunity to study the varied erosional forms, including small alcoves and honeycombing, found in the sandstone of the cliff. Here also is an interesting collection of historical inscriptions. Some of the older ones are carefully and artistically incised into the rock. A few of the oldest were written in pencil and still amazingly survive; the earliest one noted was "J. C. Peebles, 1837." Please do not be so thoughtless as to deface any of these notes from the past or add your own definitely unhistorical inscriptions. The shelf ends at a jumble of fallen rock which may be the remains of a previous version of Raven Rock (Figure 221). Make your way down this pile of rock to the base of the ledge you have just

Figure 221. A possible fallen predecessor of Raven Rock.

359

descended. Rockgrin Arch is about 20 feet to the west, back along the base of the cliff. Once there, you will see that you passed over it as you descended the rock shelf. Slide Arch is directly behind Rockgrin Arch and is best reached by carefully making your way around the end of the low rock buttress holding Rockgrin Arch.

From here, you will have to retrace your steps back to the top of Raven Rock and then down the trail to the parking lot. No matter how tempting it may be, do NOT attempt to go straight down the hillside from the arches to get back to your vehicle. The slope is extremely steep and made treacherous by fallen leaves. Worse yet, it comes to an abrupt end at the brink of the lower cliffs, a situation you will come upon without warning. Once there, you will find your retreat back up the slope extremely difficult due to its steepness and the unstable nature of its surface. It is far better to keep to the designated trail and enjoy a well-deserved supper in nearby Portsmouth than to find yourself stuck in an impossible situation hoping that by some miracle your cell phone will find a signal.

Ladd Natural Bridge (Figure 144), located near Cutler, is privately owned, but the Ladd family has generously dedicated the 35 acres surrounding it as a state nature preserve. It can be visited once an access permit is secured. A short access trail following the old lane originally built when the arch was a commercial venture leads to a sign announcing that you are entering the preserve. A few steps more bring you to the top of the natural bridge with a good view of the skylight.

Shoemaker State Nature Preserve, site of Cedar Fork Arch, has an access trail that begins at a parking lot on Pine Gap Road near Peebles, Ohio. It can be reached from the junction of State Routes 41 and 32 by going east on State Route 32 for 3.9 miles, then right on Portsmouth Road for 1 mile and right on Pine Gap Road for 0.1 mile. The preserve entrance is on the right side of the road. Access to the preserve is across private property and visitors are required to remain on the trail at all times. This also applies within the preserve which protects several rare plant species that could be irreparably harmed by thoughtless trampling. Although the arch is on the opposite side of the stream from the trail, it is readily visible (Figure 56). Mattress Arch (Figure 57), located in the right-hand abutment of the larger arch as viewed from the trail, can also be seen, but its namesake mattress has now been removed. The preserve boundary runs along the top of the arch, so most of it remains privately owned, which is another reason to remain on the trail.

Ohio Division of Forestry
2045 Morse Road, Building H-1
Columbus, OH 43229-6693
877-247-8733
http://www.dnr.state.oh.us/forestry/
Hocking State Forest: 19275 State Route 374, Rockbridge, OH 43149; 740-385-4402. Feature protected: Big Pine Creek Pillar.

A unit of Ohio DNR, the Division of Forestry (Ohio DOF) is responsible for promoting the health of Ohio's forests through education, landowner assistance and example. It administers a number of state forests for multiple public benefits including forest products, recreation, watershed protection and wildlife. All listed features so far located on state forest property have been found in Hocking State Forest whose nearly 9,700 acres, located in the popular Hocking Hills tourist region, are interspersed with state park land, nature preserves and private property. Given its location, it is not surprising that one of the main management goals of this forest is to provide a forest cover that enhances woodland recreation. In common with other state forests, it is also managed for wildlife habitat, timber production, forest research, watershed protection and demonstrations of good forest management practices. The forest contains several miles of hiking trails, bridle trails, a horsemen's camp and a rock-climbing/rappelling area. Visitors are free to walk on forest property from sunrise to sunset, although care should be taken during the hunting season. The forest closes at dark. The forest's web page, accessed through the Ohio DOF web site, provides a map of the forest. The broken nature of forest land ownership can make finding a route that avoids trespassing difficult.

The one feature that can be seen relatively easily is Big Pine Creek Pillar. The journey begins at the Hocking State Forest climbing/ rappelling area parking lot which is located on the north side of Big Pine Road, the same road leading to the Conkles Hollow State Nature

361

Preserve parking lot, about three-quarters of a mile east of its junction with State Route 374. From the parking lot, cross the road and follow the trail leading to the climbing area across a narrow metal bridge over Big Pine Creek. You will probably notice the blue blazes of the Buckeye Trail, Ohio's premier hiking adventure route, along the way. At the far side of the bridge, the foot trail crosses the bridle trail which forded the creek upstream of the bridge. Turn left, following the blazes of the Buckeye Trail onto the bridle trail which parallels the south bank of the stream. Within a quarter mile, the trail turns to ascend the hillside, eventually turning again to follow the base of a sandstone cliff upstream, weaving its way between large slump blocks. It passes two very large monolithic slabs of sandstone and then angles up a break in the cliff. The trail here can be muddy and rocky, so watch your footing. Once on the flat ground at the top of the cliff, the trail makes a wide curve and heads back downstream. As the trail begins to turn, look straight ahead. Just beyond the edge of the cliff marking the entrance of a side gorge into the main valley of Big Pine Creek you will see the top of a separated slab of rock. This is the table of Big Pine Creek Pillar. A short walk to the cliff edge will bring you to the best view of the pillar (Figure 222). Although such off-trail use is permitted on state forest land, you are expected to walk gently and be considerate of the plants and animals that call this place home. Looking down and to your left while standing above the pillar, you will see the triangular opening of an impressive gravity arch formed by a massive slump block tipped back against the cliff. From this vantage point, the 3-foot difference in elevation between the cliff rim and the top of the lower slump block is readily visible.

Ohio State Parks
> 2045 Morse Road, Building C-3
> Columbus, OH 43229-6693
> 614-265-6561
> *http://www.dnr.state.oh.us/parks/*

Hocking Hills State Park: 19852 State Route 664 South, Logan, OH 43138; 740-385-6842. Features protected: Old Mans Pantry and Rock House.

Figure 222. A view of Big Pine Creek Pillar from the top of the cliff behind it.

The mission of the Ohio State Parks (Ohio SP) is to provide outdoor recreation experiences for park visitors. Presently managing over 174,000 acres of land and water in more than 70 parks, this unit of Ohio DNR provides camping and cottage facilities, golf courses, lodges, interpretive programs, swimming beaches, boat ramps, picnic areas and over 1,100 miles of trails.

Hocking Hills State Park contains two accessible listed features and one feature that potentially could be listed. The 2,356 acres of the park include camping facilities, cottages, a dining lodge, interpretive programs and some of the most scenic trails in Ohio featuring cliffs, waterfalls and deep hemlock-shaded gorges. The popular Old Mans Cave unit is the heart of Hocking Hills State Park. Here are found the visitor's center and dining lodge. Here, too, is found one of the more easily viewed pillared-alcove arches in the area. From the parking lot, cross State Route 664 to the trail head, marked by a bulletin board display, at the top of the main hollow. Access Old Mans Cave Trail here, crossing a stone bridge over the top of Upper Falls which gives a fine view to the left of what could be called "Upper Upper Falls." The trail forks shortly after the bridge; take the branch heading down into the gorge. This will bring you to the large, circular pool at the base of Upper Falls. Continue downstream on the trail, following the energetic little stream coming out of the pool as it drops ever deeper into the gorge. About 100 yards downstream from the pool, look to the cliff on the far side of the gorge. Old Mans Pantry will appear as two dark holes in the rock, one somewhat circular and the other a long, horizontal oval (Figure 100). Please resist the temptation to leave the trail for a closer look. A short distance below, the trail crosses the stream and then crosses back. From the bridge over the second crossing you can look down into Devils Bathtub (Figure 93), an elongated pothole carved out by sand and gravel carried in the swirling currents of the stream. The trail crosses the stream once again and meets a trail entering from the right that ascends to the visitor's center by a series of steps through a short side gorge. At the junction, look up at the rim of the cliff on the opposite side of the stream where Eagle Rock juts out. Beneath this massive stone are two indentations separated by a pillar of rock. This is a potential pillared-alcove arch (Eagle Rock Arch, Figure 153) which has yet to be surveyed. Although no other arches are visible on the remainder of Gorge Trail, there are many worthwhile things to see including Sphinx Head, Old Mans Cave proper and Lower Falls.

Rock House, Ohio's largest natural arch which also happens to be a natural-tunnel has been a visitor destination for well over a century-and-a-half. A separate unit of Hocking Hills State Park, it is located 2 miles west of Gibisonville and can be reached via State Route 374 which connects all the units of the park. From the parking lot, head toward the log picnic shelter you passed as you entered the park. If you had to park in the second lot farther in, you can either walk the road back to the picnic shelter above the first lot or take the trail that begins near the rest room at the second lot. This trail will bring you to the top of the hewn rock stairs. There you can either descend the stairs and take this tour in reverse, or turn left and follow Rim Trail to its junction with the trail from the first parking lot. Near the log picnic shelter, a bulletin board, drinking fountain and "Warning, dangerous cliffs" sign mark the beginning of the trail to Rock House. A short distance down, Rim Trail meets the main trail from the right. Continue straight ahead on the main trail to another trail junction. The trail ahead makes a gentle descent to the bridge across the stream at the lowest point of the trail. The right hand trail is the original path. It is a much steeper descent, marked by large blocks of native rock installed as steps. Halfway down this trail, an overlook gives a view of the oft-photographed west end of Rock House (Frontispiece). The trail then continues down and meets the alternate trail at the footbridge below. If you are fortunate, a thread-like waterfall will be misting the head of the short side-gorge that fronts the west face of the arch.

From here, it is all uphill (Figure 223). A long series of steps carries the trail up the northern face of the cliff into which the arch has been carved, passing several of the "Gothic-arched" windows (Figure 224). One of the narrowest of these gives access to the interior of Rock House. Notice the historic carvings cut into the giant pillars on either side of the entrance, but please do not add any of your own. Once in the shaded interior of the arch, take time to allow your eyes to adjust. Notice the striking colors staining the walls. If you are fortunate enough to visit at a time when others have more important things to do, you might enjoy a moment of solitude. Take advantage of the opportunity to sit or stand quietly and listen to the musical dripping of water, remembering that this gentle noise is the voice of the force whose patient efforts carved this massive natural cathedral. A good photograph of the interior of Rock House is almost impossible to take without a tripod, so let the scene print itself on your mind.

From the arch, the trail continues along the face of the cliff, offering a view of the eastern end of the arch. Careful observers will note a small "archlet" overhead a short distance down the trail. Within a few hundred yards, the trail turns back on itself and ascends the side

Figure 223. The steps leading up to the entrance of Rock House.

Figure 224. The view through one of the "Gothic-arched" windows of Rock House.

of the gorge by means of steps hewn into the rock. At the top the trail splits. The branch leading straight ahead ends at the second parking lot. The right-hand branch is the Rim Trail and follows the edge of the cliff, but not too closely, above Rock House. It ends at the main trail. Turn left and follow this trail back to the upper parking lot where you began.

Ohio Division of Wildlife
  2045 Morse Road, Building G
  Columbus, OH 43229-6693
  800-WILDLIFE (945-3543)
  http://www.dnr.state.oh.us/wildlife/

  Features protected: Trimmer Arch, Skull Cave Arch and Woodbury Natural Bridge.

  Ohio DNR's Division of Wildlife (Ohio DOW) is charged with conserving and improving fish and wildlife resources and their habitats in the state, and promoting their use and appreciation by the public. This is the agency to go to for hunting and fishing licenses. They also produce several educational publications and assign officers to each county to handle wildlife related situations and enforce wildlife regulations. The division has acquired a number of wildlife areas and fishing lakes across the state, some of them of considerable size. Although primarily purchased for hunting and fishing, these areas are open to other kinds of minimal recreation such as hiking, bird watching and photography. Information and maps are available on the division's web site.

  Two Ohio DOW wildlife areas protect natural arches. Although they are accessible to the general public, visitors should remember that they are in hunting areas. Find out what hunting seasons are open at the time of your projected visit and be constantly aware of your surroundings while on the area so as to avoid any unpleasant encounters. It would be best to plan your trip so as not to coincide with deer season. Wearing a hat or vest in blaze orange or some other bright color will make you more visible to those hunters who might be using the site at the same time.

  Trimmer Arch is found in Paint Creek Lake Wildlife Area which is located north of US Route 50, 11 miles east of Hillsboro. The wildlife area contains over 11,000 acres along the Paint Creek and Rattle-snake Creek branches of Paint Creek Reservoir, a US Army Corps of Engineers flood control project. Land not needed for actual operation of the dam and related facilities has been leased by the Federal Government to Ohio DNR. Part of it is managed by Ohio SP as Paint Creek State Park which provides camping and boating facilities as well as picnic areas and hiking trails. The rest of the leased land is managed by Ohio DOW.

Two accessible arches are located on the east side of the Paint Creek branch of the reservoir within the wildlife area. They can be reached by taking Rapid Forge Road north from US Route 50 to Weller Lane. Turn west onto Weller Lane and follow it to its end at a paved turn-around for school buses. Park as far to one side of the lot as possible in order to leave room for the school bus. The rest of the journey will be made on foot, partly through open, scrub-filled old-fields and partly through young woodlands. A hat, sunscreen and plenty of water are highly recommended.

From the parking area, follow the old road bed across a narrow stream. The bridge has been removed, requiring a scramble down a short slope to the stream and then up the opposite bank and back onto the old road. Within a matter of feet the old road turns to the left (east) and passes through a tree-lined fence-row running north-south. This fence-row marks the eastern boundary of public land; once the old road passes through, it enters private property. Do not follow the road beyond this point for, not only will you be trespassing, but you will also have difficulty finding the natural arches, and finding them is already difficult enough. Instead, make your way across the reverting farm field ahead of you, paralleling the line of trees and keeping it on your left. Study this tree line; it will become an important landmark later.

There are no maintained trails in this part of the wildlife area. Such trails as you may find depend for their existence on how frequently they are used and where the deer happen to want to go that year. However, a trail heading generally south from the curve of the old road and paralleling the tree line has been in existence for at least fifteen years and appears to be fairly permanent. It will be your best guide through the bramble patches of the old field. Just be sure to keep the tree-marked fence line on your left so you do not move off-course by turning onto intersecting trails. At the far side of the old-field, the trail, which may be quite faint by now, leads into a young woodland in a shallow valley where it meets up with an old farm lane and turns to follow it west, downstream. Within a short distance the trail turns back to the south and crosses a small stream in the bottom of the valley. If you are on course, you will see a very low ledge of exposed rock and a 2-foot-high waterfall at the stream crossing. The trail then leads up and over a divide covered with more brush and young trees. Watch closely for the trail as it may nearly disappear in places.

Eventually, you will descend into a larger valley with a noticeably flat floor. As you approach the creek in its bottom, you will see outcrops of Peebles Dolomite on its banks straight ahead and to your right, downstream. At the creek, turn around and take a good look at where you just came from so you will know where to leave this valley on your return. After committing the scene to memory, turn and make your way up the stream. Very shortly you will come to a fork. You will want to follow the left (north) branch which comes down through a dolomite outcrop, creating an interesting little amphitheater with a low waterfall at its head. While it may be possible to scale the unstable slope beside the waterfall, it may not be a wise thing to do, so, after exploring the weathered paleokarst of the amphitheater, retrace your steps and climb up the north side of the valley to skirt the rock outcrop. Follow the north branch up a few hundred yards and you will come to a second fork in the stream. Trimmer Arch is located in this junction (Figure 225). Follow the north (left) fork of this second branching to Skull Cave Arch, found a few hundred feet upstream (Figure 68). To see the sinkhole openings of this arch you will have to scramble up the steep slope beside the stream entrance. The sinkhole will be visible once you top the crest (Figure 226). Be advised that the boundary line of the wildlife area runs very near to the sink-hole, so do not wander beyond it. From here you can either retrace your steps to Trimmer Arch or make your way south across a minor divide to the next small stream and follow it down for a view

Figure 225. Close view of Trimmer Arch.

of the south side of Trimmer Arch. From here you can return to the parking lot, making good use of that mental picture you formed earlier.

Woodbury Natural Bridge is found in Woodbury Wildlife Area, located south of the town of Warsaw in Coshocton County. Originally containing fewer than 2,000 acres, the wildlife area was later enlarged to over 19,000 acres by the purchase of a large tract of mostly strip-mined land. The natural bridge is located on the original wildlife area and so the vegetation surrounding it has had time to mature. Take State Route 60 south for 2.5 miles from its junction with US Route 36 in the village of Warsaw to County Road 17. Turn left and take the county road southeast 1 mile to Township Road 56, also known as Woodbury Lane, which comes in from the right where County Road

Figure 226. Skull Cave Arch showing the double sinkhole into which the arch opens. The main opening is in the right side of the larger, more distant sinkhole. The "window" opens into the smaller sinkhole in front.

17 makes a sharp turn to the left. There may not be a road sign at this junction, but an Ohio DOW sign will indicate that you are entering the wildlife area. Turn west (right) onto Woodbury Lane and follow it 1.5 miles past a pond and a primitive campground to a turn-around at the end of the road. Leave the car here, being careful not to block the turn-around, and follow the abandoned roadbed on foot up the narrowing stream valley for about 400 feet to a tributary stream coming in from the right. At this point you will still be within sight of your vehicle at the turn-around. Follow the tributary stream up toward its head until you come to the natural bridge (Figure 133). The bridge is neither high nor spectacular, but is very interesting for having formed

371

in an isolated exposure of limestone surrounded by sandstone bedrock. It is also unmistakable since the stream you are following flows beneath it.

Ohio Historical Society
    1982 Velma Avenue
    Columbus, OH 43211
    800-686-6124; 614-297-2300
    http://www.ohiohistory.org
    Fort Hill State Memorial, 13614 Fort Hill Road, Hillsboro, OH 45133. Features protected: Natural Y Arch, The Keyhole and Spring Creek Arch.

The Ohio Historical Society, a unique public/private partnership, was incorporated in 1885 for the purpose of preserving and interpreting the archaeological and historical resources of Ohio. In addition to important collections of historical material, the society maintains a system of historic sites across the state including prehistoric mounds, forts and locations associated with people important in state and national history. Several memorials also contain valuable examples of Ohio's natural heritage. Only Fort Hill State Memorial contains natural arches, but the collection found here is the most dense concentration of these features known from a single site in the state. Located on Township Road 256 off State Route 41 in Highland County 5 miles north of Sinking Springs, this 1,200-acre area protects one of the best preserved Hopewell hilltop enclosures in the country. It also contains most of the impressive gorge of Baker Fork. With 11 miles of hiking trails and a picnic area, the memorial makes an excellent day trip for those seeking a wild corner of Ohio. It is open year-round except during deer hunting season in the fall. For hours and other information, check the Fort Hill page on the society's web site.

The seven natural arches and one pillar in the memorial are all found in the depths of Baker Fork Gorge. Three of the most interesting arches are visible from the trail. The other features are remote, difficult to access and require an off-trail use permit to visit. The best times to see the arches are in early spring before the trees have fully leafed out, and in the fall after their leaves have fallen. Winter is also a good time for viewing the arches, but be aware that ice and snow can make the steep trails very treacherous and the memorial may not be open for visitation. Spring has the added attraction of a spectacular wildflower show with whole hillsides full of Ohio's state wildflower, the Large-flowered Trillium (*Trillium grandiflorum*). Plan on spending the better part of a day on the trails, for there is much more to see than just the natural arches. There will be many vertical stretches to negotiate including some stairs, so visitors should be in good physical condition. The topographic map in Figure 40 indicates the topography involved and the general location of the trails. The following itinerary will take you to all three of the easily seen arches.

Begin at the end of the parking lot nearest the red brick museum. The trail starts in a broad mowed area where the access road meets the parking lot. A sign post marks the beginning of both Gorge Trail and Deer Trail. Blazes of light blue paint show that Deer Trail is also part of the Buckeye Trail's 1,444-mile loop around the entire state. Part of Deer Trail also is combined with North Country National Scenic Trail, running from New York to North Dakota, and American Discovery Trail, which crosses the country from Delaware to California, so hikers on Fort Hill's Deer Trail are participating in a national experience.

Your route for this trip to Fort Hill's arches, however, is Gorge Trail which is about 4.1 miles long. The trail soon leaves the easy flat ground and follows Baker Fork into the hills, climbing and falling as it crosses several ravines dropping into Baker Fork. The mature forest in this section contains some good-sized trees and one of the best spring wildflower displays in the memorial. Enjoy them here because the climbs and descents of this section of the trail will be constantly repeated ahead, only on a larger scale and you might find yourself too much out of breath at times to appreciate the scenery. Within a short distance, the first low knob of Peebles Dolomite appears, jutting out into the creek. Beyond it the trail crosses over an even higher dolomite knob, providing hikers their first close look at this rock layer which contains most of the natural arches in the region.

Shortly after passing this knob, the trail reaches a junction. Deer Trail, along with Buckeye Trail and the two national trails, turns right to cross Baker Fork and ascends the steep hillside beyond, passing over Reeds Hill and Jarnigan Knob and then descending to Baker Fork to rejoin Gorge Trail. Stay on Gorge Trail which turns left at the junction, leading to a log cabin shelter in a mowed clearing visible a short distance ahead (Figure 227). The trail re-enters the woods to the right of the shelter and continues its lonely way along the east side of Baker Fork, climbing up and over ever-higher knobs of Peebles Dolomite projecting out from the side of Fort Hill. The cliffs on either side of the Baker Fork gradually close in until they have constricted the gorge to its narrowest point. This is most likely the site of the col that was breeched by rising waters of the glacial meltwater lake which once filled the level area from the parking lot all the way into Beech Flats north of the memorial.

Just beyond this point, the trail rises to the top of a noticeably high knob of rock, passing close to a precipitous drop into the creek before making a sharp bend to the left and descending a long set of wooden steps into Beech Ravine. Pause here to look toward the far

Figure 227. The log cabin shelter alongside Gorge Trail in Fort Hill State Memorial.

side of the gorge. Directly across from your vantage point a long, dark opening appears. This is the front of Natural Y Arch (Figure 228). The double skylight is not usually visible from this point, but it is possible to get an idea of the size of the arch and its setting in the upper part of the Peebles Dolomite.

Continuing down the stairs and up and over several intervening knobs, the trail eventually comes once again to the level of Baker Fork. The Keyhole, one of Ohio's most impressive natural arches, is readily visible on the far side of the stream (Figure 39). The tall, narrow opening of the arch clearly exhibits the characteristic shape of an enlarged vertical crevice. Due to the sensitivity of the area, the opening may not be accessed. However, it is possible to reach the top of the arch by turning onto the combined Deer and Buckeye trails which here rejoin Gorge Trail. They will lead to a ford a short distance downstream. Once across Baker Fork, the combined trails ascend the steep side of the gorge and then turn to cross the top of The Keyhole. From here you can look down into the small, enclosed valley

Figure 228. Natural Y Arch as seen from Gorge Trail in Fort Hill State Memorial.

through which Bridge Creek flows before entering The Keyhole. The steep shale slopes on either side of the trail can be treacherous, so please do not leave the marked path.

Return to Baker Fork and continue downstream on Gorge Trail. A short distance from The Keyhole, a small gravity arch through which Sulphur Creek flows can be seen on the left (east) side of the trail. Its opening frames a view of a low waterfall just upstream (Figure 28).

Farther downstream, the trail rises onto a wide, flat bench marking the top of the Peebles Dolomite high above Baker Fork. It follows this bench to a final set of steps leading down to Spring Creek where the trail turns left (east) up a narrow gorge. Spring Creek Arch sits on the opposite side of the stream a short distance up from Baker Fork (Figure 43). The entrance and skylight can both be seen from the trail, and the rough, pitted nature of the weathered Peebles Dolomite is readily visible. The talus slope beneath the entrance supports a carpet of Large-flowered Trillium which would be irreparably damaged by foot traffic, so please limit your viewing sites to the legal trail.

From here you have two options for returning to the parking lot. You can turn around and retrace your steps to see if the arches look any different coming from the opposite direction, or you can continue on Gorge Trail which circles the back side of Fort Hill before dropping down to where your vehicle waits. The second option has rewards of its own. Should you choose it, follow Gorge Trail up to the head of Spring Creek Gorge where it meets an unpaved service road.

At this point, the trail has climbed from the Peebles Dolomite onto the overlying Ohio Shale, moving forward in geologic time from the Silurian Period to the Devonian Period. The shale, which forms the greater part of the hills in this region, appears as small, broken, flat brownish-gray to black plates littering the trail. Where the overlying soil and plant cover have fallen away, thinly laminated blocks of shale can be seen in place, looking like the stacked pages of an immense book. Shale is more easily eroded than dolomite and forms a slope rather than a cliff. Noting how thin and brittle the layers are, one can easily see why any arches or pillars formed in it must of necessity be small and short-lived.

At the head of the creek, the trail turns left and follows the service road up the south side of Fort Hill. Deer Trail splits off from Gorge Trail after a short distance to cross the saddle between Fort Hill and Easton Hill and then descends to the parking lot. This lower-level trail is best for those who have had enough climbing, even though it is not by any means level. It has the added benefit of passing through some impressive forested coves at the heads of small streams draining the sides of the hills.

More ambitious hikers can continue on Gorge Trail which ascends Fort Hill and traverses its back side at a higher level, eventually descending to the parking lot. However, having come this far, it would be a pity to leave without seeing Fort Hill itself, the feature for which the area was originally acquired. Follow Gorge Trail to its junction with Fort Trail. Turn left onto Fort Trail and follow it around the south end of the summit of Fort Hill. On this level stretch, the shale remnants littering the trail lower down have been replaced by pieces of brown sandstone, showing that you have passed onto the Berea Sandstone of Mississippian age which caps the hill. Within a short distance the trail turns abruptly to the right and ascends the steep slope formed by the resistant sandstone. Near its top you finally pass through the low walls of prehistoric Fort Hill (Figure 229). At this point, the walls of the "fort" are below the actual top of the hill and so the trail continues up until it reaches the flat mesa-like summit 423 feet above Baker Fork.

From here, the trail parallels the inner side of the western wall of the earthwork. The embankment is 1.6 miles long and completely encircles the level top of the hill. It is made of earth and stone, most of which evidently came from the ditch paralleling the inside of the wall. Early settlers assumed the enclosure, which contains about 40 acres, had been built for defense, hence the name Fort Hill. However, the thirty-three openings in the wall and the interior ditch which averages 50 feet in width would make it difficult to defend. Today the embankment is believed to have been built by people of the Hopewell culture for ceremonial purposes between 100 B.C. and A.D. 500.

Figure 229. The low walls of the Hopewellian Fort Hill enclosure, showing one of the many gaps that would make defending it difficult.

In addition to several other hilltop "forts," including Fort Ancient in Warren County, the Hopewell people also built a series of large geometric earthworks in the wide bottomlands at the junction of Paint Creek and the Scioto River to the northeast, some of which are now protected as Hopewell Culture National Historical Park. As you walk through this sacred space of a vanished culture, contemplate the devotion that would inspire people to raise such a work using only digging sticks, woven baskets and muscles. Remember this: the hike up the hill was no easier for them than for you!

At the far end of Fort Hill the trail reaches an overlook with a long view out over Beech Flats, the now-dry bottom of the glacial lake responsible for forming Baker Fork Gorge. Here the trail turns to the right, passes out between the walls of the fort and then makes a steep descent of the hillside, bringing you back to the parking lot and hopefully a well-deserved drink of something cold.

Marietta Unit, Wayne National Forest

27750 State Route 7
Marietta, OH 45750
740-373-9055
http://www.fs.fed.us/r9/wayne/

Features protected: Irish Run Natural Bridge and Fern Pillar.

The boundaries of Ohio's only national forest enclose over 800,000 acres, but much of this land remains privately owned. Split into three units, the national forest is a patchwork quilt of federal property interspersed with private and state-owned land. Even so, it is still able to demonstrate responsible timber management while providing opportunities for hunting, fishing, hiking, camping and picnicking. For the most part, the forest is second and third growth, ranging in age from young to fairly mature. Oaks and hickories dominate natural woodlands with hemlock and several species of pine found in some places. There are also many large areas of non-native trees planted to rehabilitate former strip mines which have been acquired as part of the forest. Only two listed features are found on the forest, but one of them is definitely impressive.

Irish Run Natural Bridge, located on the Marietta Unit of the national forest, can be reached by taking either State Route 26 or State Route 7 east from Marietta. Route 26, which a fellow arch aficionado has called "a connection of several curves called a road," follows the winding course of the Little Muskingum River. It has been designated by the US Forest Service as the Covered Bridge Scenic Byway, and several picturesque examples of these historic structures are visible from the road. At a distance of 26 miles from its junction with Route 7 on the eastern side of Marietta, Route 26 meets State Route 260. Turn right (east) and go 3.5 miles to its junction with Township Road 34 which enters from the right (south) in the village of Hohman. State Route 7, the alternate way to this point, has fewer curves, but is no less scenic for it closely traces the north bank of the Ohio River and is part of the Ohio River Scenic Byway. It also has the advantage of an exit from Interstate 77 at Marietta and is less prone to flooding than is State Route 26.

From the junction of Route 7 with Route 26 in Marietta, drive east 29 miles to its junction with State Route 260 in the town of New Matamoras. Turn left (west) and follow Route 260 for 7 miles to its junction with Township Road 34 which enters from the left (south) in the village of Hohman.

Regardless of which direction you take to reach Hohman, you will find little to indicate that you have arrived. There is no town as such. There is only a house and a galvanized road-sign post, which might not even carry a sign, in the center of the triangular intersection of State Route 260 and Township Road 34. A little over a mile down the township road, just beyond an old cemetery on the left, another dirt road, marked by a national forest directional sign announcing a North Country Trail parking area, enters from the left. Turn and follow this road a short distance to its end at a larger cemetery. Park here, being careful not to block the two jeep trails meeting the road across from the cemetery entrance.

The rest of the journey will be made on foot. Walk back to the jeep trails which have a common junction with the road. Ignore the gated trail straight ahead, turning instead to the rutted trail leading downhill to the left which takes you below the edge of the upland on which the cemetery and your car sit. A short distance down you will see an oil well head with a walking beam pump (Figure 230). This area is part of an active well field, although its production is limited. Do not approach the equipment; these wells are private property and the pumps can start working without warning.

About 750 feet down the jeep trail, a foot trail enters on the left, marked by a directional sign for the natural bridge and Great Cave. Follow this trail 50 feet downhill to the head of a narrow ravine which opens to the right. A trail on the near side of the ravine leads to the natural bridge; it is marked by a sign post, although the sign may be missing. Before taking this trail, follow the main trail about 75 feet, across the head of the ravine and then down its far side for a view of Great Cave (Figure 231). Actually a large alcove shelter typical of this part of the state, Great Cave gives a preview of Irish Run Natural Bridge since it has formed in the

380

Figure 230. The oil well pump on the way to Irish Run Natural Bridge.

Figure 231. Great Cave.

same rock strata and in the same manner as the alcove which gave rise to the bridge.

Return to the trail on the near side of the ravine and follow it for about a quarter mile. Although the ravine deepens in this direction, the trail maintains a level course so that eventually it is high above the small stream below. Two noticeable rocks will appear downslope to the left. The smaller one is Fern Pillar (Figure 192), the other is its neighboring standing stone (Figure 232). A short distance beyond them a sign for the natural bridge points down a side trail to the left. You are now only a few steps from the skylight of Irish Run Natural Bridge (Figure 233). Use extreme caution as you descend this short trail as the drop into the alcove behind the arch is very steep and one careless step could leave you stranded in a very wild place. Several pictures and more trail information can be found on the Wayne National Forest web site.

Figure 232. The standing stone next to Fern Pillar.

Figure 233. The skylight of Irish Run Natural Bridge as seen from above.

Vinton Furnace Experimental Forest
c/o Northeastern Research Station
359 Main Road
Delaware, OH 43015
740-596-4238
http://www.fs.fed.us/ne/delaware/4153/vfef.html
Feature protected: Arch Rock.

Arch Rock is located within Vinton Furnace Experimental Forest, a 3,247-acre recovering forest which is part of the nearly 16,000-acre Raccoon Ecological Management Area. The forest, named for the remains of a nineteenth-century iron furnace located on the grounds, is used by researchers to develop information on woodland management practices related to timber production, recreation and wildlife. During operation of the iron furnace, the forest was repeatedly cut off to provide fuel. Later it was grazed. The middle-aged woodland now covering the hills has grown up since the land was purchased for timber production. Many of the research projects undertaken here relate to the management and improvement of the oak-dominated ecosystem which occupied much of southern Ohio at the time of Euro-American settlement.

383

Vinton Furnace operated between 1854 and 1883 and produced nearly 6,000 tons of iron per year. In the beginning, it used charcoal made from trees cut on company land. Every year, 350 acres of woodland was cut to supply the required charcoal. The furnace converted to coal in 1868. Although the furnace itself is now an unstable pile of rubble, a battery of 24 brick Belgian-coke ovens are still in reasonably good shape. They were used to heat the coal, driving out impurities to create coke which is a more efficient fuel than wood for making iron. Built in 1875, they are thought to be the only ones of their kind in the United States. The area surrounding the furnace has been set aside as a sanctuary.

Visitors interested in seeing Arch Rock should call the Experimental Forest far enough in advance to allow the staff to send a map showing the trail to the arch. The area is open for hunting and so visiting during hunting season is not recommended. Getting to the forest is an adventure in itself. From the town of McArthur, take State Route 93 south to the village of Dundas. Once there, turn left on State Route 324, cross the railroad tracks, drive 0.2 miles and turn left at the Vinton Furnace Experimental Forest directional sign. Continue another 4.3 miles to the first intersection. Continue straight on Township Road 7 a distance of 0.4 miles to a parking lot at an orange gate. Signs to the forest are posted at all intersections. Should all else fail, just follow the power lines.

From the parking lot at the gate, it is a walk of several miles to the arch, so relax and enjoy the hike. On the way, you may pass evidence of research being conducted. Please do not disturb anything you find; such thoughtless action could compromise or even destroy the value of these experiments. Arch Rock is not large, but it is scenic (Figure 138). If you choose your weather carefully, you can have a pleasant day in one of the largest continuous tracts of forest to be found in Ohio.

Franz Theodore Stone Laboratory
P. O. Box 119
Put-in-Bay, OH 43456
419-285-1800; 614-292-8949
http://ohioseagrant.osu.edu/stonelab
Feature protected: Needles Eye.

Stone Lab, created in 1895, is the nation's oldest freshwater biological field station. This island campus of The Ohio State University includes the Lake Erie research facility of the Ohio Sea Grant College Program and offers courses on both the undergraduate and graduate level as well as classes for select high school students, classroom teachers and the public. Located on six-acre Gibraltar Island in the harbor of Put-in-Bay, the facility is dominated by the three-story laboratory building and Cooke Castle, the 1865 mansion of Jay Cooke, industrialist and Civil War financier. It can be reached by water taxi from Put-in-Bay on South Bass Island, which itself can be reached only by boat or airplane. A seasonal ferry service to South Bass Island is available.

On the northeast end of Gibraltar Island lies Needles Eye, one of Ohio's few active coastal arches. Although the opening is only 3 feet wide, it can be easily viewed from the seawall in front of Perrys Victory and International Peace Memorial on South Bass Island. Walk northeast along the seawall, watching the rocky end of Gibraltar Island opposite the laboratory building until a sliver of light breaks through. This is Needles Eye. A good pair of binoculars will help. Getting closer requires a boat, although care must be taken to avoid rocks and shoals. The island is owned by the university and permission is required to access it; the best views of the arch, however, are from the open water of the lake (Figure 75). Those wishing to see it from the land side can watch the laboratory web site for notice of public open houses during which tours are given of Gibraltar Island and the facilities located there.

The Miami County Park District
2645 East State Route 41
Troy, OH 45373
937-335-MCPD (6273)
http://www.miamicountyparks.com/
Feature protected: Greenville Falls Arch.

With over 1,900 acres of natural areas under its care, the Miami County Park District in western Ohio offers waterfalls, prairies, hiking trails and many more interesting things to explore. Among these treasures is Greenville Falls State Nature Preserve. Although owned by Ohio DNAP as part of its program to protect the Stillwater State Scenic River System, the 79-acre preserve is leased to the Miami County Park District for management purposes. Over the past several years, the district has improved the trail system, adding interpretive signs and an overlook at the falls.

Located west of the town of Covington, Greenville Falls can be reached most easily by following US Route 36 west out of town about a mile to its junction with Range Line Road. Take Range Line Road south (left) to Covington-Gettysburg Road and turn west (right). The entrance to the preserve parking lot is a few hundred yards west of the intersection.

From the parking lot, a trail leads toward the river, passing through an old-field which is being allowed to revert to native growth. Watch for patches of remnant prairie plants such as Gray-headed Coneflower (*Ratibida pinnata*) as you pass through. The trail leads to an overlook built above Greenville Falls, actually more of a cascade than a waterfall (Figure 234). Its 20-foot drop is impressive, nonetheless. From the overlook you can also see the remains of an old water-powered electric generating station on the south side of the gorge below the falls.

Figure 234. Greenville Falls as seen from the overlook in Greenville Falls State Nature Preserve.

Two trails lead from the overlook. The one heading upstream follows the north bank of the river, passing the remains of the wooden dam that once channeled the river's flow into the generating station. The other trail follows the river downstream from the falls, keeping a safe distance from the rim of the gorge. About 100 yards from the falls overlook, this trail passes the top of Greenville Falls Arch. An interpretive plaque explains the arch, but the feature itself is not readily visible from this vantage point. The trail continues past the "hanging fen" downstream from the arch where uncommon native Ohio wetland plants such as Shrubby Cinquefoil (*Potentilla fruticosa*) grow on the cliff face, surviving on the cold groundwater seeping out through the rock. The trail then curves back through the reverting old-field to the parking lot.

For a better view of the arch, return to the junction of Covington-Gettysburg and Range Line roads, and turn south (right). As you cross the bridge over Greenville Creek, you will catch a glimpse of the gorge and the falls upstream. Just beyond the bridge is the entrance to another parking area on the west (right) side of the road. This recent addition to the park was funded by an Ohio DNR Nature Works Grant. The land was purchased from the Dayton Power and Light Company which still operates an electricity transmitting facility, the last vestige of the old generating station, on the site. A short trail leads to an overlook built on the foundation of the generating station where interpretive signs explain its operation. A side trail leads down stone steps into the gorge, ending on the bank of Greenville Creek. Although built mainly for the use of fishermen, this trail also gives visitors their best view of Greenville Falls Arch, which appears as a dark, shadowed recess in the cliff wall on the opposite side of the river directly across from the bottom of the steps (Figure 72).

Glen Helen Ecology Institute
405 Corry Street
Yellow Springs, OH 45387-1895
937-769-1902
http://www.antioch-college.edu/glenhelen/
Feature protected: Pompeys Pillar.

This 1,000 acre nature preserve on the east edge of the village of Yellow Springs in Greene County had its origin in a gift of land to Antioch College made in 1929 by Hugh Taylor Birch to honor his daughter Helen. Twenty-five miles of trails lead visitors past dolomite cliffs, waterfalls, four-century-old trees and Yellow Spring for which its home town is named. Trailside Science Museum is the center for scheduled hikes and programs in The Glen, as it is commonly called. The nearby Glen Helen Building contains The Nature Shop, offering hundreds of nature and ecology-oriented products, including field guides. Visitors can also see the Raptor Center where injured birds of prey are rehabilitated for eventual release back into the wild. An on-site Outdoor Education Center has provided residential environmental education experiences for school children for over 50 years. This scenic area, the downstream neighbor to John Bryan State Park, is most easily reached from US Route 68 via Corry Street which meets the federal route at the north end of the village business district. The main parking lot for The Glen is located less than one-quarter mile south on Corry Street.

The impressive dolomite erosional remnant known as Pompeys Pillar is found on the east side of Yellow Springs Creek a short distance downstream from Yellow Spring. From the main parking lot on Corry Street, take the trail past the Trailside Science Museum down into the gorge and across Yellow Springs Creek where it dead-ends at a trail paralleling the creek. Turn left (upstream) on this trail. Pompeys Pillar can be seen between the trail and the line of cliffs on the right side a short distance from the junction (Figure 167).

Camp Christopher
1930 North Hametown Road
Bath, OH 44333
330-376-2267
http://www.akroncyo.org
Feature protected: Camp Christopher Natural Bridge.

The staff and campers of Camp Christopher are justifiably proud of their natural bridge (Figure 126) and enjoy showing it off to interested visitors. Those wishing to visit Camp Christopher Natural Bridge should contact the camp before arriving to make the proper arrangements. Since this is an active youth camp, it will be necessary to schedule a time when a staff member or camper will be available to serve as a guide to the arch.

Your walk will take you into the conglomerate "ledges" that form the impressive backdrop of the camp. Once on their top, you will quickly arrive at the skylight of the natural bridge. The most accessible view of the opening is from the skylight since the front of the arch opens into a very narrow, but scenic, miniature gorge. If you are fortunate to have a guide not pressed by other duties, he or she might take the time to show you some of the other interesting sites found in the ledges, including an interesting cabin ruin at a scenic overlook.

Olentangy Indian Caverns

1779 Home Road

Delaware, OH 43015

740-548-7917

http://www.olentangyindiancaverns.com

Features protected: Olentangy Caverns Arch and Leatherlips Arch.

Olentangy Indian Caverns holds the only underground arches yet listed in Ohio. Olentangy Caverns Arch (Figure 82) is found in the Indian Council Chamber, and Leatherlips Arch (Figure 83) opens behind the Indian Lovers Bench in Echo Chamber. Olentangy Caverns is not large and is readily accessible to those who are able to handle enclosed places and a long flight of stairs. In addition to the arches, the caverns has a small display of flowstone and other cave formations. For the most part, its passageways are good examples of the widening of vertical crevices, a process involved in the formation of several above-ground arches in Ohio. The underground arches found here, however, are a result of the differential solution of the limestone bedrock and are therefore considered to be bedrock-texture-remnant arches. The cave is located on Home Road which meets State Route 23 south of Delaware, Ohio. It is open daily from April 1 through October 31, 9:30 A.M. to 5:00 P.M. An entry fee is charged.

Epilogue

Adventure is where you find it, and sometimes it is closer than one might think. What began as a short-term casual project to locate and record Ohio's natural arches soon evolved into a grand adventure that rarely took me more than an easy half-day's drive from my own familiar back yard. In the process, I found an Ohio richer and more fascinating than I could have imagined. Beyond our expanding cities and wide farm fields there still exist echoes of an earlier time when the forces of Nature reigned over all and the works of Man were almost inconsequential.

These remnants of the natural world are an inheritance beyond price. They are the seedbeds of future greatness, a savings account protecting immense possibilities, a responsibility to be handed on to future generations. Some of this tattered former world remains by design; much of it still exists only through the vagaries of good fortune. Each generation decides the fate of the natural heritage it has inherited.

Fortunately, wise minds in past years have recognized the value of this natural heritage and fought to protect parts of it. The parks, nature preserves, wildlife areas and managed forests we enjoy today are a result of their perseverance and hard work. But if the story of The Devils Tea Table in Morgan County teaches us anything, it is that this heritage is only one misguided, uncontested decision away from destruction. It behooves each of us to be ever vigilant in its protection. It is, after all, not our inheritance alone, but that of our children and those who will come after them as well. They will be allowed to enjoy only so much of it as we deign to bequeath to them. If we destroy our children's inheritance for short-term selfish gain, then we will deserve all the opprobrium future generations will heap upon our heads (Figure 235).

We do have a choice. We also have a responsibility. The agencies which manage our natural places are chronically underfunded and overworked; support their efforts. Opportunities to acquire wild places not yet secured constantly arise; encourage their protection. Decisions which affect the health and well-being of the natural world we inhabit are made every day; let your voice be heard. Take the time to seek out what is left of the original Ohio — stand in silence and let the wind speak to you; take time to truly see the animals, the plants and, yes, even the geology that inhabits such places in all their wondrous complexity. What you come to know, you will appreciate. What you appreciate, you will come to love. What you love, you will fight for — and if ever anything needed champions in this present day, it is Ohio's natural heritage. Champions have risen in the past. They can rise again. You can be one of them. Future generations gazing in awe at Ohio's natural arches and pillars will thank you.

Figure 235. Visitors of a past generation enjoy Rockbridge. Will we allow the next generation to do the same?

Appendixes

Appendix I

A Tabulated List of Ohio's Documented Natural Arches

NUMBER	NAME	LOCATION	SPAN
OH-A-ADA-01	Riverbend Arch	Adams County	12 feet (3.6 meters)
OH-A-ADA-02	Roadside Arch	Adams County	4.5 feet (1.4 meters)
OH-A-ADA-03	Small Roadside Arch	Adams County	3.25 feet (1 meter)
OH-A-ADA-04	Smaller Roadside Arch	Adams County	3 feet (0.9 meter)
OH-A-ADA-05	Scioto Brush Creek Arch	Adams County	25 feet (7.6 meters)
OH-A-ADA-06*	Cedar Fork Arch	Adams County	6.5 feet (2 meters)
OH-A-ADA-07*	Mattress Arch	Adams County	3 feet (0.9 meter)
OH-A-ADA-08	Crawl Arch	Adams County	9 feet (2.7 meters)
OH-A-ADA-09	Blocked Arch	Adams County	7 feet (2.1 meters)
OH-A-ADA-10	Bundle Run Arch	Adams County	24 feet (7.3 meters)
OH-A-ADA-11	Tiffin Arch	Adams County	12 feet (3.7 meters)
OH-A-ADA-12	Castlegate Arch	Adams County	11 feet (3.3 meters)
OH-A-ADA-13	Ohioview Arch	Adams County	7.3 feet (2.2 meters)
OH-A-ADA-14	Old Womans Kitchen, West Arch	Adams County	4.7 feet (1.4 meters)
OH-A-ADA-15	Old Womans Kitchen, East Arch	Adams County	6.75 feet (2.1 meters)
OH-A-ADA-16	Covered Arch	Adams County	5.1 feet (1.5 meters)
OH-A-ATH-01	Mineral Arch	Athens County	25 feet (7.6 meters)
OH-A-ATH-02	Mustapha Natural Bridge	Athens County	35 feet (10.7 meters)
OH-A-CLA-01	Tecumseh Arch	Clark County	2.7 feet (0.8 meter)

CLEARANCE	CATEGORY	FORMATION
24 feet (7.3 meters)	Vertical crevice enlargement	Peebles Dolomite
3.5 feet (1.1 meters)	Bedrock texture remnant	Peebles Dolomite
1.5 feet (0.5 meter)	Bedrock texture remnant	Peebles Dolomite
2 feet (0.6 meter)	Bedrock texture remnant	Peebles Dolomite
8 feet (2.4 meters)	Bedrock texture remnant	Peebles Dolomite
14 feet (4.3 meters)	Bedrock texture remnant	Peebles Dolomite
2 feet (0.6 meter)	Bedrock texture remnant	Peebles Dolomite
5 feet (1.5 meters)	Bedrock texture remnant	Peebles Dolomite
5 feet (1.5 meters)	Bedrock texture remnant	Peebles Dolomite
5 feet (1.5 meters)	Joint bedding plane enlargement	Peebles Dolomite
4 feet (1.2 meters)	Cave collapse	Peebles Dolomite
9.6 feet (2.9 meters)	Bedrock texture remnant	Peebles Dolomite
1.4 feet (0.4 meter)	Pillared alcove	Bisher Formation
2 feet (0.9 meter)	Breeched alcove	Peebles Dolomite
4 feet (1.2 meters)	Breeched alcove	Peebles Dolomite
2.3 feet (0.7 meter)	Bedrock texture remnant	Peebles Dolomite
5.5 feet (1.7 meters)	Breeched alcove (vertical crevice)	Pottsville Group (?) Sandstone
11 feet (3.3 meters)	Breeched alcove (vertical crevice)	Hockingport Sandstone Lentil
3.5 feet (1.1 meters)	Bedding plane enlargement	Cedarville and Springfield dolomites

Appendix 1, continued

NUMBER	NAME	LOCATION	SPAN
OH-A-COL-01	Beaver Creek Natural Bridge	Columbiana County	8 feet (2.4 meters)
OH-A-COS-01*	Woodbury Natural Bridge	Coshocton County	5 feet (1.5 meters)
OH-A-DEL-01*	Olentangy Caverns Arch	Delaware County	2.25 feet (0.7 meter)
OH-A-DEL-02*	Leatherlips Arch	Delaware County	1.6 feet (0.5 meter)
OH-A-FAI-01	Hintz Hollow Arch	Fairfield County	3.3 feet (1 meter)
OH-A-FRA-01	Dublin Arch	Franklin County	8 feet (2.4 meters)
OH-A-GEA-01	Cats Den Natural Bridge	Geauga County	6.7 feet (2 meters)
OH-A-GEA-02	Bass Lake Natural Bridge	Geauga County	13 feet (4 meters)
OH-A-GRE-01	Low Arch	Greene County	10 feet (3 meters)
OH-A-HIG-01	Boundary Arch	Highland County	5.7 feet (1.3 meters)
OH-A-HIG-02*	Natural Y Arch	Highland County	29 feet (8.8 meters)
OH-A-HIG-03*	The Keyhole	Highland County	2 feet (0.6 meter)
OH-A-HIG-04	Baker Fork Arch	Highland County	15.7 feet (4.8 meters)
OH-A-HIG-05*	Spring Creek Arch	Highland County	18 feet (5.5 meters)
OH-A-HIG-06	The Passage	Highland County	7 feet (2.1 meters)
OH-A-HIG-07	Big Cave Natural Bridge	Highland County	8 feet (2.4 meters)
OH-A–HIG-08	Hidden Arch	Highland County	3.7 feet (1.1 meters)
OH-A-HIG-09*	Miller Natural Bridge	Highland County	46 feet (14 meters)
OH-A-HIG-10*	Miller Arch	Highland County	9.5 feet (2.9 meters)
OH-A-HIG-11	Jawbone Arch North	Highland County	8.5 feet (2.6 meters)
OH-A-HIG-12	Jawbone Arch South	Highland County	4.6 feet (1.4 meters)
OH-A-HOC-01*	Rockbridge	Hocking County	92 feet (28 meters)
OH-A-HOC-02	Balcony Natural Bridge	Hocking County	11 feet (3.3 meters)
OH-A-HOC-03	Brineinger Hollow Natural Bridge	Hocking County	16.75 feet (5.1 meters)
OH-A-HOC-04	Chapel Ridge Natural Bridge	Hocking County	12 feet (3.7 meters)

CLEARANCE	CATEGORY	FORMATION
3.7 feet (1.1 meters)	Breeched alcove (vertical crevice)	Lower Freeport Sandstone
2.2 feet (0.7 meter)	Breeched alcove (vertical crevice)	Putnam Hill Limestone
3 feet (0.9 meter)	Bedrock texture remnant	Delaware Limestone
5.25 feet (1.6 meters)	Bedrock texture remnant	Delaware Limestone
1.5 feet (0.5 meter)	Bedrock texture remnant	Black Hand Sandstone
6.75 feet (2 meters)	Cave collapse	Columbus Limestone
8.5 feet (2.6 meters)	Vertical crevice enlargement	Sharon Formation
4 feet (1.2 meters)	Breeched alcove (waterfall)	Sharon Formation
30 inches (0.8 meter)	Bedding plane enlargement	Cedarville and Springfield dolomites
9 feet (2.7 meters)	Vertical crevice enlargement	Peebles Dolomite
5 feet (1.5 meters)	Bedrock texture remnant	Peebles Dolomite
35 feet (10.7 meters)	Vertical crevice enlargement	Peebles Dolomite
8 feet (2.4 meters)	Bedrock texture remnant	Peebles Dolomite
5 feet (0.2 meter)	Bedrock texture remnant	Peebles Dolomite
2 feet (0.6 meter)	Bedrock texture remnant	Peebles Dolomite
4 feet (1.2 meters)	Joint bedding plane enlargement	Peebles Dolomite
2.2 feet (0.7 meter)	Bedrock texture remnant	Peebles Dolomite
3.5 feet (1.1 meters)	Joint bedding plane enlargement	Peebles Dolomite
5 feet (1.5 meters)	Bedrock texture remnant	Peebles Dolomite
4.7 feet (1.4 meters)	Joint bedding plane enlargement	Peebles Dolomite
3.3 feet (1 meter)	Joint bedding plane enlargement	Peebles Dolomite
40 feet (12.2 meters)	Breeched alcove	Black Hand Sandstone
17.5 feet (5.5 meters)	Breeched alcove	Black Hand Sandstone
3.3 feet (1 meter)	Breeched alcove	Black Hand Sandstone
4 feet (1.2 meters)	Breeched alcove	Black Hand Sandstone

Appendix 1, continued

NUMBER	NAME	LOCATION	SPAN
OH-A-HOC-05	Conkles Hollow Arch	Hocking County	10 feet (3 meters)
OH-A-HOC-06	Polypody Arch	Hocking County	8.75 feet (2.7 meters)
OH-A-HOC-07	Hagley Hollow Arch	Hocking County	26.5 feet (8.1 meters)
OH-A-HOC-08	Unger Hollow Natural Bridge	Hocking County	24 feet (7.3 meters)
OH-A-HOC-09*	Old Mans Pantry	Hocking County	20 feet (6.1 meters)
OH-A-HOC-10*	Rock House	Hocking County	185 feet (56 meters)
OH-A-HOC-11	M Arch	Hocking County	18 feet (5.5 meters)
OH-A-HOC-12	1811 Arch	Hocking County	
OH-A-HOC-13	Early Arch	Hocking County	31 feet (9.4 meters)
OH-A-HOC-14*	Saltpetre Cave Arch	Hocking County	11.8 feet (3.6 meters)
OH-A-HOC -15*	Surprise Arch	Hocking County	15.5 feet (4.7 meters)
OH-A-HOC-16	Muddy Crack Arch	Hocking County	1.2 feet (0.4 meter)
OH-A-HOC-17	Three Hole Arch	Hocking County	3.5 feet (1.1 meters)
OH-A-HOC-18	Annex Arch	Hocking County	5.6 feet (1.7 meters)
OH-A-JAC-01	Hole-in-the-Wall Arch	Jackson County	3.9 feet (1.2 meters)
OH-A-JAC-02	Lake Katharine Arch	Jackson County	15 feet (4.6 meters)
OH-A-JAC-03	Peephole Arch	Jackson County	3 feet (0.9 meter)
OH-A-LOR-01	Amherst Arch North	Lorain County	4 feet (1.2 meters)
OH-A-LOR-02	Amherst Arch South	Lorain County	3 feet (1.2 meters)
OH-A-MIA-01*	Greenville Falls Arch	Miami County	22 feet (3.3 meters)
OH-A-MIA-02	Cobble Arch	Miami County	3.6 feet (1.1 meters)
OH-A-MON-01	David Road Natural Bridge	Montgomery County	4.5 feet (1.4 meters)
OH-A-MRG-01	Lucas Run Natural Bridge	Morgan County	27 feet (8.2 meters)
OH-A-OTT-01	Needles Eye	Ottawa County	3 feet (0.9 meter)
OH-A-OTT-02	LeMarin Arch	Ottawa County	4.5 feet (1.4 meters)
OH-A-OTT-03	North Sugar Rock Arch	Ottawa County	2.5 feet (0.8 meter)
OH-A-OTT-04	South Sugar Rock Arch	Ottawa County	12 feet (3.7 meters)

CLEARANCE	CATEGORY	FORMATION
8 feet (2.4 meters)	Bedrock texture remnant	Black Hand Sandstone
3.7 feet (1.1 meters)	Bedrock texture remnant	Black Hand Sandstone
7 feet (2.1 meters)	Bedrock texture remnant	Black Hand Sandstone
6 feet (1.8 meters)	Breeched alcove	Black Hand Sandstone
3.7 feet (1.1 meters)	Bedrock texture remnant	Black Hand Sandstone
25 feet (7.6 meters)	Vertical crevice enlargement	Black Hand Sandstone
6 feet (1.8 meters)	Bedrock texture remnant	Black Hand Sandstone
	Bedrock texture remnant	Black Hand Sandstone
6.7 feet (2 meters)	Bedrock texture remnant	Black Hand Sandstone
7.5 feet (2.3 meters)	Bedrock texture remnant	Black Hand Sandstone
1.9 feet (0.6 meter)	Breeched alcove	Black Hand Sandstone
8.6 feet (2.6 meters)	Vertical crevice enlargement	Black Hand Sandstone
1.8 feet (0.5 meter)	Bedrock texture remnant	Black Hand Sandstone
2.6 feet (0.8 meter)	Bedrock texture remnant	Black Hand Sandstone
1.8 feet (0.5 meter)	Breeched alcove	Sharon Formation
2 feet (0.6 meter)	Breeched alcove	Sharon Formation
1.8 feet (0.5 meter)	Bedrock texture remnant	Sharon Formation
2.5 feet (0.8 meter)	Broken bedding planes	Berea Sandstone
3.2 feet (1 meter)	Broken bedding planes	Berea Sandstone
4 feet (1.2 meters)	Joint bedding plane enlargement	Cedarville Dolomite
1.2 feet (0.4 meter)	Bedrock texture remnant	Pleistocene gravel
3 feet (0.9 meter)	Breeched alcove (?)	Cemented Pleistocene gravel
11 feet (3.3 meters)	Breeched alcove (vertical crevice)	Monongahela (?) Group Sandstone
9.5 feet (2.7 meters)	Vertical crevice enlargement	Put-in-Bay Dolomite
6 feet (1.8 meters)	Pillared alcove	Tymochtee Dolomite
4 feet (1.2 meters)	Vertical crevice enlargement	Tymochtee Dolomite
6 feet (1.8 meters)	Vertical crevice enlargement	Tymochtee Dolomite

Appendix 1, continued

NUMBER	NAME	LOCATION	SPAN
OH-A-PIK-01	Lions Den Arch	Pike County	9.5 feet (2.9 meters)
OH-A-PIK-02	Pike Arch	Pike County	30 feet (9.1 meters)
OH-A-ROS-01*	Trimmer Arch	Ross County	14 feet (4.6 meters)
OH-A-ROS-02*	Skull Cave Arch	Ross County	7.9 feet (2.4 meters)
OH-A-ROS-03	Vertical Dome Arch	Ross County	1 foot (0.3 meter)
OH-A-SCI-01*	Raven Rock Arch	Scioto County	15 feet (4.6 meters)
OH-A-SCI-02*	Rockgrin Arch	Scioto County	5 feet (1.5 meters)
OH-A-SCI-03*	Slide Arch	Scioto County	3.6 feet (1.1 meters)
OH-A-SCI-O4	The Penthouse	Scioto County	
OH-A-SUM-01*	Camp Christopher Natural Bridge	Summit County	5 feet (1.5 meters)
OH-A-VIN-01*	Arch Rock	Vinton County	9.7 feet (3 meters)
OH-A-WAS-01	Liberty Natural Bridge	Washington County	6 feet (1.8 meters)
OH-A-WAS-02*	Ladd (Big) Natural Bridge	Washington County	40 feet (12.2 meters)
OH-A-WAS-03	Little Natural Bridge	Washington County	44 feet (13.4 meters)
OH-A-WAS-04*	Irish Run Natural Bridge	Washington County	22 feet (6.7 meters)

* A publicly accessible site

CLEARANCE	CATEGORY	FORMATION
4.75 feet (1.4 meters)	Bedrock texture remnant	Monroe Limestone
10 feet (3 meters)	Headward erosion	Sharon Formation
8.6 feet (2.6 meters)	Bedding plane enlargement	Greenfield Dolomite
3.6 feet (1.1 meters)	Cave collapse	Greenfield Dolomite
4 feet (1.2 meters)	Vertical crevice enlargement	Peebles Dolomite
7 feet (2.1 meters)	Bedrock texture remnant	Mississippian Sandstone
1.25 feet (0.4 meter)	Bedrock texture remnant	Mississippian Sandstone
5 feet (1.5 meters)	Bedrock texture remnant	Mississippian Sandstone
	Bedrock texture remnant	Mississippian Sandstone
20 feet (6.1 meters)	Vertical crevice enlargement	Sharon Formation
8.25 feet (2.5 meters)		Conemaugh Group Sandstone
5 feet (1.5 meters)	Breeched alcove (vertical crevice and collapse)	Sewickley (?) Sandstone
17 feet (5.2 meters)	Breeched alcove (vertical crevice)	Hockingport Sandstone Lentil
26 feet (7.9 meters)	Breeched alcove (vertical crevice)	Hockingport Sandstone Lentil
5 feet (0.2 meter)	Breeched alcove (vertical crevice and collapse)	Dunkard Group Sandstone

Appendix II

A Tabulated List of Ohio's Documented Natural Pillars

NUMBER	NAME	LOCATION	HEIGHT
OH-P-ADA-01	Teakettle Rock	Adams County	16.5 feet (5 meters)
OH-P-ADA-02	Wood Rat Tower	Adams County	23 feet (7 meters)
OH-P-ADA-03	Prow Tower	Adams County	19 feet (5.8 meters)
OH-P-ADA-04	Cute Tower	Adams County	8 feet (2.4 meters)
OH-P-ATH-01	Mineral Tea Table	Athens County	10.5 feet (3.2 meters)
OH-P-GRE-01*	Pompeys Pillar	Greene County	15 feet (4.6 meters)
OH-P-GRE-02	John Bryans Window	Greene County	7 feet (2.1 meters)
OH-P-HIG-01	Fort Hill Tea Table	Highland County	12 feet (3.7 meters)
OH-P-HOC-01*	Big Pine Creek Pillar	Hocking County	27 feet (8.2 meters)
OH-P-HOC-02	Balanced Rock	Hocking County	15 feet (4.6 meters)
OH-P-HOC-03	Flat Iron Rock	Hocking County	15 feet (4.6 meters)
OH-P-MRG-01	The Devils Tea Table	Morgan County	25 feet (7.6 meters)
OH-P-PIK-01	Chimney Rock	Pike County	23 feet (7 meters)
OH-P-PIK-02	Little Chimney Rock	Pike County	11 feet (3.3 meters)
OH-P-PIK-03	Morel Tea Table	Pike County	20 feet (6.1 meters)
OH-P-WAS-01	Dale Tea Table	Washington County	ca. 20 feet (6.1 meters)
OH-P-WAS-02	Dunbar Tea Table	Washington County	23 feet (7 meters)
OH-P-WAS-03	Fern Pillar	Washington County	15 feet (4.6 meters)

* A publicly accessible site

WIDTH	CATEGORY	FORMATION
7 feet (2.1 meters)	Tea table	Peebles Dolomite
10 feet (3 meters)	Monolithic rock	Peebles Dolomite
ca. 3 feet (0.9 meter)	Monolithic rock	Peebles Dolomite
2 feet (0.6 meter)	Chimney	Peebles Dolomite
ca. 5 feet (1.5 meters)	Tea table	Lower Freeport Sandstone
	Tea table	Cedarville and Springfield dolomites
	Tea table	Cedarville and Springfield dolomites
6 feet (1.8 meters)	Tea table	Peebles Dolomite
3 feet (0.9 meter)	Tea table	Black Hand Sandstone
3 feet (0.9 meter)	Tea table	Black Hand Sandstone
1.5 feet (0.5 meter)	Tea table	Black Hand Sandstone
5 feet (1.5 meters)	Destroyed	Dunkard Group (?) Sandstone and Shale
15 feet (4.6 meters)	Chimney	Sharon Formation
5 feet (1.5 meters)	Chimney	Sharon Formation
ca. 6 feet (1.8 meters)	Tea table	Sharon Formation
ca. 3 feet (0.9 meter)	Tea table	Dunkard Group Sandstone
5 feet (1.5 meters)	Tea table	Dunkard Group Sandstone
9 feet (2.7 meters)	Chimney	Dunkard Group Sandstone

Appendix III

English-Metric Conversion Tables

English-Metric Conversions

1 inch	2.54 centimeters
1 foot	0.3048 meter
1 yard	0.9144 meter
1 mile	1.6093 kilometers
1 acre	0.4047 hectare

Metric-English Conversions

1 centimeter	0.3937 inch
1 meter	3.2808 feet
1 kilometer	0.621 mile
1 hectare	2.471 acres

Bibliography

Addington, Arch R. 1927. The Litten Natural Bridges and Closely Associated Phenomena, Eastern Owen County, Indiana. *Proceedings of the Indiana Academy of Science* 37: 143–151.

Andrews, E. B. 1871. Report of Labors in the Second Geological District, During the Year 1870. Pp. 55–251 *in Geological Survey of Ohio Report of Progress in 1870*. Columbus, OH: Nevins & Myers, State Printers.

————. 1873a. Report on Athens County. *Report of the Geological Survey of Ohio* 1(1): 261–293.

————. 1873b. Report on Morgan County. *Report of the Geological Survey of Ohio* 1(1): 294–313.

————. 1874a. Report on the Geology of Washington County. *Report of the Geological Survey of Ohio* 2(1): 453–508.

————. 1874b. Surface Geology. *Report of the Geological Survey of Ohio* 2(1): 441–452.

Anonymous. n.d. *Nelson-Kennedy Ledges State Park* (Brochure). Columbus, OH: Ohio Department of Natural Resources.

————. n.d. *Ohio Caverns: Where Nature Carved a Fairyland.* West Liberty, OH: Ohio Caverns, Inc.

————. 1975. Devil's Tea Table was natural curiosity of valley. *The Marietta Times.* May 24.

Ausich, William I. 1981. *The Regional Paleontology and Stratigraphy of Ohio: Report Number PA8103WA.* Dayton, OH: Wright State University Department of Geological Sciences.

————. 1987. John Bryan State Park, Ohio: Silurian Stratigraphy. Pp. 419–422 *in* D. L. Biggs (ed.), *Geological Society of America Centennial Field Guide — North-Central Section.* Boulder, CO: Geological Society of America.

Baker, Jack. 1965. *Drainage History of a Part of the Hocking River Valley.* Field Trip Guide. Columbus, OH: The Ohio Academy of Science.

Baker, Ronald L. 1982. *Hoosier Folk Legends.* Bloomington, IN: Indiana University Press.

Baker, Stanley W., and Jason Watkins. 2005. *Phase I Archaeological Survey for the ATH-124-1.90 (CR 62) Highway Relocation and Reconstruction Project Located in Troy Township, Athens County, Ohio (PID 79122)*. Columbus, OH: Cultural Resources Unit, Office of Environmental Services, Ohio Department of Transportation.

Barnes, F. A. 1987. *Canyon Country Arches and Bridges*. Moab, Utah: Canyon Country Publications.

Barnett, V. H. 1908. A natural bridge due to stream meandering. *Journal of Geology* 16: 73–75.

Beckedahl, H. R. 1987. Rock mass strength determinations as an aid to landscape interpretation. Pp. 393–397 in V. Gardiner (ed.), *International Geomorphology 1986, Part I*. Chichester, NY: John Wiley & Sons, Ltd.

Birot, Pierre. 1968. *The Cycle of Erosion in Different Climates*. Berkeley, CA: University of California Press.

Blatt, Harvey, Gerard Middleton, and Raymond Murray. 1972. *Origin of Sedimentary Rocks*. Englewood Cliffs, NJ: Prentice-Hall, Inc.

Bownocker, J. A. 1920. *Geologic Map of Ohio*. Columbus, OH: Ohio Department of Natural Resources, Division of Geological Survey.

Braun, E. Lucy. 1928. *The Vegetation of the Mineral Springs Region of Adams County, Ohio*. Columbus, OH: The Ohio State University Bulletin 32(30).

————. 1969. *An Ecological Survey of the Vegetation of Fort Hill State Memorial, Highland County, Ohio and Annotated List of Vascular Plants*. Columbus, OH: Ohio Biological Survey, New Series Bulletin 3(3).

Bray, Edmund C. 1985. *Ancient Valleys, Modern Rivers: What the Glaciers Did*. Saint Paul, MN: The Science Museum of Minnesota.

Bretz, J. Harlen. 1942. Vadose and phreatic features of limestone caverns. *Journal of Geology* 50: 675–811.

————. 1956. *Caves of Missouri*. Rolla, MO: Missouri Division of Geological Survey and Water Resources.

Caldwell, J. A. 1880. *History of Belmont and Jefferson Counties, Ohio*. Wheeling, WV: Historical Publishing Company.

Camp, Mark J. 2006. *Roadside Geology of Ohio*. Missoula, MT: Mountain Press Publishing Company.

Carman, J. Ernest. 1972. *The Geologic Interpretation of Scenic Features in Ohio*. Columbus, OH: Ohio Division of Geological Survey Reprint Series No. 3.

Carter, Charles H., Donald E. Guy, and Jonathan A. Fuller. 1981. Coastal geomorphology and geology of the Ohio shore of Lake Erie. Pp. 433–456 in Thomas G. Roberts (ed.), *GSA Cincinnati '81 Field Trip Guidebooks*. Alexandria, VA: American Geological Institute.

Cleland, Herdman F. 1910. North American natural bridges, with a discussion of their origin. *Bulletin of the Geological Society of America* 21: 313–338.

Cole, W. Storrs. 1934. Identification of erosion surfaces in eastern and southern Ohio. *Journal of Geology* 42(1): 285–294.

Coffey, George N. 1958. Major glacial drainage changes in Ohio. *The Ohio Journal of Science* 58:43–49.

———. 1961. Major Preglacial, Nebraskan and Kansan glacial drainages in Ohio, Indiana, and Illinois. *The Ohio Journal of Science* 61:295–313.

Collins, Horace R. 1979. *The Mississippian and Pennsylvanian (Carboniferous) Systems in the United States — Ohio.* United States Geological Survey Professional Paper 1110-E. Washington, DC: United States Government Printing Office.

Collins, Horace R., and Bradley E. Smith. 1977. *Geology and Mineral Resources of Washington County, Ohio.* Ohio Division of Geological Survey Bulletin 66. Columbus, OH: Ohio Division of Geological Survey.

Coogan, A. H. 1996. Ohio's surface rocks and sediments. Pp. 31-50 *in* Rodney M. Feldmann (ed.), *Fossils of Ohio.* Ohio Division of Geological Survey Bulletin 70. Columbus, OH: Ohio Division of Geological Survey.

Coogan, A. H., R. M. Feldmann, E. J. Szmuc, and V. Mrakovich. 1974. Sedimentary environments of the Lower Pennsylvanian Sharon Conglomerate near Akron, Ohio. Pp. 19–41 *in* R. A. Heimlich and R. M. Feldmann (eds.), *Selected Field Trips in Northeastern Ohio.* Ohio Division of Geological Survey Guidebook No. 2. Columbus, OH: Ohio Division of Geological Survey.

Coogan, A. H., Richard A. Heimlich, Robert J. Malcuit, Kennard B. Bork, and Thomas L. Lewis. 1981. Early Mississippian deltaic sedimentation in central and northeastern Ohio. Pp. 113–152 in Thomas G. Roberts (ed.), *GSA Cincinnati '81 Field Trip Guidebooks.* Boulder, CO: American Geological Institute.

Corgan, James X., and John T. Parks. 1979. *Natural Bridges of Tennessee.* Nashville, TN: Tennessee Division of Geology Bulletin 80.

Cotton, C. A. 1942. *Climatic Accidents in Landscape Making.* New York, NY: John Wiley and Sons, Inc.

Cruikshank, Kenneth M., and Atilla Aydin. 1994. Role of fracture localization in arch formation, Arches National Park, Utah. *Geological Society of America Bulletin* 106:879–891. Boulder, CO: Geological Society of America.

Cummings, Byron. 1910. The great natural bridges of Utah. *National Geographic Magazine* 21:157–167.

Dean, Stuart L., Byron R. Kulander, Jane L. Forsyth, and Ronald M. Topton. 1991. Field guide to joint patterns and geomorphological features of northern Ohio. *The Ohio Journal of Science* 91(1): 2–15.

DeLong, Richard M. 1967. *Bedrock Geology of the South Bloomingville Quadrangle, Hocking and Vinton Counties, Ohio.* Ohio Division of Geological Survey Report of Investigations No. 63. Columbus, OH: Ohio Division of Geological Survey.

Division of Natural Areas and Preserves. 1996 (with 1998 and 2000 update pages). *Directory of Ohio's State Nature Preserves.* Columbus, OH: Ohio Division of Natural Areas and Preserves.

Dougherty, Percy H. (ed.). 1985. *Caves and Karst of Kentucky.* Kentucky Geological Survey Special Publication 12, Series 11. Lexington, KY: University of Kentucky.

Durrell, Lucile, and Richard Durrell. 1975. *The Wilderness: The Charles A. Eulett Preserve, Adams County, Ohio*. Cincinnati, OH: The Cincinnati Museum of Natural History.

———. 1979. Today's landscape. Pp. 48–57 in Michael B. Lafferty (ed.), *Ohio's Natural Heritage*. Columbus, OH: The Ohio Academy of Science.

Durrell, Richard H. 1977. A recycled landscape. *Quarterly of the Cincinnati Museum of Natural History* 14(2). (Reprinted 1982 as a booklet)

Farnham, Barbara, and Alexander Farnham. 1988. *Kingwood Township of Yesteryear.* Stockton, NJ: Kingwood Studio Publications.

Fellows, L. D. 1965. Cutters and pinnacles in Greene County, Missouri. *The National Speleological Society Bulletin* 27: 143–150. Huntsville, AL: The National Speleological Society.

Floto, Bernard A. 1955. *The Possible Presence of Buried Niagaran Reefs in Ohio and Their Relationship to the Newburg Oil and Gas Zone*. Ohio Division of Geological Survey Report of Investigation No. 24. (Reprinted 1962)

Foerste, August F. 1915. *Geology of Dayton and Vicinity*. Indianapolis, IN: The Hollenbeck Press.

Foos, Annabelle M. (ed.). 2003. *Pennsylvanian Sharon Formation, Past and Present: Sedimentology, Hydrogeology, and Historical and Environmental Significance*. Ohio Division of Geological Survey Guidebook No. 18. Columbus, OH: Ohio Division of Geological Survey.

Ford, D. C. 1978. Threshold and limit effects in karst geomorphology. Pp. 345–362 *in* D. C. Coates and J. D. Vitek (eds.), *Thresholds in Geomorphology*. London, UK: George Allen and Unwin.

Forsyth, Jane L. 1959. *The Beach Ridges of Northern Ohio*. Ohio Division of Geological Survey Information Circular No. 25. Columbus, OH: Ohio Division of Geological Survey.

———. 1961. *Dating Ohio's Glaciers*. Ohio Division of Geological Survey Information Circular No. 30. Columbus, OH: Ohio Division of Geological Survey.

———. 1962. Upland flats in eastern Ohio — peneplain or one-cycle erosion surface? *The Ohio Journal of Science* 62(6): 311–314. Columbus, OH: The Ohio Academy of Science.

———. 1965. Geology's contribution to Ohio's landscapes. *The American Biology Teacher* 27(5): 358–362.

———. 1973. Late-glacial and postglacial history of western Lake Erie. *The Compass* (Sigma Gamma Epsilon) 51(1): 16–26.

Fuller, J. Osborn. 1955. Source of Sharon Conglomerate of northeastern Ohio. *Geological Society of America Bulletin* 66: 159–176.

Galloway, William Albert. 1932. *The History of Glen Helen*. Columbus, OH: F. J. Heer Ptg. Co. (Reprinted 1977, Knightstown, IN: The Bookmark)

Gerrard, A. J. 1988. *Rocks and Landforms*. London, UK: Unwin Hyman.

Gray, Henry H. 1991. Origin and history of the Teays drainage system: The view from midstream. Pp. 43–50 *in* Wilton N. Melhorn and John P. Kempton (eds.),

Geology and Hydrogeology of the Teays-Mahomet Bedrock Valley System. Geological Society of America Special Paper 258. Boulder, CO: Geological Society of America.

Gregory, H. E. 1917. *Geology of the Navajo Country*. Washington, DC: United States Geological Survey Professional Paper 93. Washington, DC: US Government Printing Office.

Gregory, H. E. 1956. *A Geologic and Geographic Sketch of Zion National Park*. Springville, UT: Art City Publishing Company.

Goldthwait, Richard P., George W. White, and Jane L. Forsyth. 1961. *Glacial Map of Ohio*. United States Geological Survey: Miscellaneous Geologic Investigations, Map I-316. Washington, DC: US Government Printing Office.

Guralnik, David B., (ed.). 1976. *Webster's New World Dictionary of the English Language*. Cleveland, OH: William Collins + World Publishing Co., Inc.

Hall, John F. 1961. *The Geology of Hocking State Park*. Ohio Division of Geological Survey Information Circular No. 8. Columbus, OH: Ohio Division of Geological Survey.

Hansen, Michael C. 1975. *Geology of the Hocking Hills State Park Region*. Columbus, OH: Ohio Division of Geological Survey Guidebook No. 4. Columbus, OH: Ohio Division of Geological Survey.

———. 1988a. Natural bridges in Ohio. *Ohio Geology* Summer: 1, 3–6.

———. 1988b. Ohio natural bridges. *Earth Science* 41(4): 10–12.

———. 1991. Campbell Hill — Ohio's summit. *Ohio Geology* Winter: 1, 3–5.

———. 1993. Natural bridges. *Timeline* 10(6): 22–29.

———. 2001. The geology of Ohio — the Mississippian. *Ohio Geology* 2001 (2): 1, 3–7.

Happ, Stafford. 1934. Drainage history of southeastern Ohio and adjacent West Virginia. *Journal of Geology* 42: 264–284.

Harrison, Joseph T. 1922. The pillars of Harrison County. *Ohio Archaeological and Historical Publications* 31: 120–127.

Hartley, Robert P. 1962. Relation of shore and nearshore bottom features to rock structure along Lake Erie. *The Ohio Journal of Science* 62: 125–131.

Hattin, Donald E., and J. Robert Dodd. 1992. *Mississippian Paleosols, Paleokarst, and Eolian Carbonates in Indiana*. Ohio Division of Geological Survey, Miscellaneous Report No. 3. Columbus, OH: Ohio Division of Geological Survey.

Hawk, Harold H. n.d. In Washington County, Natural Bridge is an Ohio Wonder. Typescript copy in files of Ohio Division of Natural Areas and Preserves, Columbus, OH.

Hinds, N. A. E. 1943. *Geomorphology*. New York, NY: Prentice-Hall, Inc.

Hobbs, H. H., III. 1982. The Rocky Fork caves (Seven Caves). *Pholeos* 3(1): 5–22.

———. 1985. Small "jamas" and "spiljas" in Kentucky and Ohio. *Pholeos* 5(1): 5–7.

Hobbs, H. H., III, and Michael F. Flynn. 1981. Selected Ohio caves. *Pholeos* 1(1 & 2): 15–24.

Hobbs, H. H., III, Naomi D. Mitchell, and Todd L. Zimmerman. 1986. A description of additional lilliputian caves formed in Ohio dolomites and limestones. *Pholeos* 7(2):6–13.

Hodge, J. T. 1878. Report on the geology of Coshocton County. *Report of the Geological Survey of Ohio* 3(1): 562–595.

Hoffman, John F. 1985. *Arches National Park: An Illustrated Guide.* San Diego, CA: Western Recreational Publications.

Holmes, Arthur. 1965. *Principles of Physical Geology.* New York, NY: The Ronald Press Company.

Horowitz, Daniel. 1993. Definition of a natural arch and related types of natural rock openings. *SPAN: Newsletter of the Natural Arch and Bridge Society* 5(3): 5–6.

Horvath, Allan L., and Dale Sparling. 1967. *Guide to the Forty-Second Annual Field Conference of the Section of Geology of the Ohio Academy of Science: Silurian Geology of Western Ohio.* Dayton, OH: University of Dayton.

Howe, Henry. 1896. *Historical Collections of Ohio in Two Volumes.* Norwalk, OH: The Laning Printing Co.

Howell, J. V. (ed.). 1960. *Glossary of Geology and Related Sciences.* Washington, DC: American Geological Institute.

Hubbard, G. D. 1924. Dimensions of the Cincinnati Anticline. *The Ohio Journal of Science.* 24(3): 161–168.

Hubbard, G. D., C. R. Stauffer, J. A. Bownocker, C. S. Prosser, and E. R. Cumings. 1915. *Geologic Atlas of the United States: Columbus Folio.* United States Geological Survey, Folio 197 — Field Edition. Washington, DC: United States Geological Survey.

Hull, Dennis N. 1990. *Generalized Column of Bedrock Units in Ohio.* Columbus, OH: Ohio Division of Geological Survey.

Hussey, John. 1878. Report of the geology of Miami County. *Report of the Geological Survey of Ohio* 3(1): 468–481.

Hyde, Jesse E. 1953. *Mississippian Formations of Central and Southern Ohio.* Ohio Division of Geological Survey Bulletin 51. Columbus, OH: Ohio Division of Geological Survey.

Interstate Publishing Co. 1883. *History of Hocking Valley, Ohio.* Chicago, IL: Interstate Publishing Co.

Indiana Geological Survey. Leaflet GM-35. Bloomington, IN: Indiana Department of Natural Resources.

Jankowski, Laurence J. 1982. The geology of Hocking Hills parks. *Ohio Woodlands* Winter: 33–37.

Janssen, Raymond E. 1943. Nature's bridges. *The Scientific Monthly* 57: 210–219. Washington, DC: American Association for the Advancement of Science.

Jefferson, Thomas. 1954 (1787). *Notes on the State of Virginia.* Chapel Hill, NC: The University of North Carolina Press.

Jennings, J. N. 1971. *Karst.* Cambridge, MA: The MIT Press.

Kahle, Charles F. 1987. Surface and subsurface paleokarst, Silurian Lockport, and Peebles Dolomites, Western Ohio. Pp. 229–255 *in* N. P. James and P. W. Choquette (eds.), *Paleokarst*. New York, NY: Springer-Verlag.

Kahle, Charles F., and Jack C. Floyd. 1972. *Geology of Silurian Rocks, Northwestern Ohio: Field Trip Guidebook for First Annual Meeting, Eastern Section American Association of Petroleum Geologists*. Columbus, OH: Ohio Geological Society.

Keller, Tom. 1984. Cedar Fork Cave. *Pholeos* 4(2): 3–4.

Kincaid, Dan. 1986. Irish Run Natural Bridge. *Wayne National Forest Newsletter*, 8.

Kind, Thomas C. 1976. The development of cross-valley natural bridges in southeastern Ohio. Pp. 13–20 in Phillip D. Phillips (ed.), *1976 Proceedings, Geography Section, Kentucky Academy of Science*. Lexington, KY: Kentucky Academy of Science.

Knoop, Paul. 1989. Ohio's natural rock formations: A bridge to the past. *Dayton Daily News*, October 14.

Krumbein, W. C., and L. L. Stoss. 1951. *Stratigraphy and Sedimentation.* San Francisco, CA: W. H. Freeman and Company.

Lamborn, Raymond E. 1951. *Limestones of Eastern Ohio*. Ohio Division of Geological Survey, Fourth Series Bulletin 49. Columbus, OH: Ohio Division of Geological Survey.

———. 1954. *Geology of Coshocton County*. Ohio Division of Geological Survey Bulletin 53. Columbus, OH: Ohio Division of Geological Survey.

Langlois, Thomas H. 1965. The waves of Lake Erie at South Bass Island. *The Ohio Journal of Science* 65: 351.

Leverett, F. 1902. *Glacial Formations and Drainage Features of the Erie and Ohio Basins*. United States Geological Survey Monograph 41. Washington, DC: United States Geological Survey.

Leverett, F., and Frank B. Taylor. 1915. *The Pleistocene of Indiana and Michigan and the History of the Great Lakes*. United States Geological Survey Monograph 53. Washington, DC: United States Geological Survey.

Lobeck, A. K. 1939. *Geomorphology: An Introduction to the Study of Landscapes*. New York, NY: McGraw-Hill Book Company, Inc.

Lohman, S. W. 1975. *Geologic Story of Arches National Park*. United States Geological Survey Bulletin 1393. Washington, DC: United States Geological Survey.

Lord, Richard C. 1923. The Black Hand Formation in north central Ohio. *The Ohio Journal of Science* 23: 124–129.

Luther, Warren Phillips. 1988. Some interesting Ohio caves in noncarbonate rocks, parts 1 and 2. *Pholeos* 9(1): 5–17.

———. 1989. Some interesting Ohio caves in noncarbonate rocks, part 3. *Pholeos* 9(2): 2–7.

Lynch, Larry. 1974. The formation of caves in the Sharon Conglomerate in Geauga County, Ohio. *Cleve-O-Grotto News* 20(6): 51–59.

Malcuit, Robert J., and Kennard B. Bork. 1987. Black Hand Gorge State Nature Preserve: Lower Mississippian deltaic deposits in east-central Ohio. Pp. 411–414 in D. L. Biggs (ed.), *Geological Society of America Centennial Field Guide — North-Central Section.* Boulder, CO: Geological Society of America.

Malott, Clyde A., and Robert R. Shrock. 1930. Origin and development of Natural Bridge, Virginia. *American Journal of Science — Fifth Series* 19: 257–273.

Martin, Wayne D., and Bernard R. Henniger. 1969. Mather and Hockingport sandstone lentils (Pennsylvanian and Permian) of Dunkard Basin, Pennsylvania, West Virginia and Ohio. *The American Association of Petroleum Geologists Bulletin* 53: 279–298.

Mattes, Merrill J. 1978 (1955). Chimney Rock on the Oregon Trail. *Nebraska History* 26. (Reprinted 1978 as a booklet)

McFarlan, Arthur C. 1954. *Geology of the Natural Bridge State Park Area.* Kentucky Geological Survey Series 9, Special Publication 4. Lexington, KY: Kentucky Geological Survey.

McGrain, Preston. 1979. *Recognition of Lapies-Type Features in the Kentucky Karst — An Example of Applied Geomorphology.* Kentucky Geological Survey Series 11, Reprint 3. Lexington, KY: Kentucky Geological Survey.

———. 1983. *The Geologic Story of Kentucky.* Kentucky Geological Survey Series 11, Special Publication 8. Lexington, KY: Kentucky Geological Survey.

McKechnie, Jean L. (editorial supervisor). 1979. *Webster's New Universal Unabridged Dictionary (Deluxe Second Edition).* New York, NY: New World Dictionaries/Simon and Schuster.

McMahan, Diana. 1983. Hike through forest, discover natural span. *The Parkersburg News,* May 22: 46.

Melhorn, Wilton N. 1992. Mansfield Natural Bridge, Parke County: Requiem for a vanished Indiana landmark. *Proceedings of the Indiana Academy of Science* 101: 179–186.

Merrill, William M. 1953. Pleistocene history of a part of the Hocking River Valley, Ohio. *The Ohio Journal of Science* 53: 143–158.

Miller, Arthur M. 1898. Natural arches of Kentucky. *Science* 7: 845–846.

Minshall, F. W. 1888. *The History and Development of the Macksburg Oil Field.* Ohio Geological Survey 6: 443–475.

Morgan, Richard G., and Edward S. Thomas. 1950. *Fort Hill, Ohio.* Columbus, OH: The Ohio Historical Society.

Murphy, J. L. 1975a. *An Archeological History of the Hocking Valley.* Athens, OH: Ohio University Press.

———. 1975b. Ohio's natural rock bridges. *The Explorer* 17(4): 15–18.

———. 1977. Ohio's natural rock bridges. *The Columbus Dispatch Magazine,* December 4.

———. 2004. Archaeological potential of standing stones in eastern Ohio. *Ohio Archaeologist* 54(3): 18–21. Lancaster, OH: The Ohio Archaeological Society.

Nadon, Gregory C., Elizabeth H. Gierlowski-Kordesch, and Joseph P. Smith. 1998. *Sedimentology and Provenance of Carboniferous and Permian Rocks of Athens County, Southeastern Ohio.* Ohio Geological Survey, Guidebook No. 15. Columbus, OH: Ohio Division of Geological Survey.

Naylor, James Ball. 1960. The Devil's Tea Table. *Morgan County Herald.* February 4.

Nevin, Charles Merrick. 1949. *Principles of Structural Geology.* New York, NY: John Wiley & Sons, Inc.

Newberry, J. S. 1873a. Geological relations of Ohio. *Report of the Geological Survey of Ohio* 1(1): 50–88.

———. 1873b. Physical geography. *Report of the Geological Survey of Ohio* 1(1): 16–49.

———. 1874. Geology of Erie County and the islands. *Report of the Geological Survey of Ohio* 2(1): 183–205.

———. 1878a. Geology of Columbiana County. *Report of the Geological Survey of Ohio* 3(1): 90–132.

———. 1878b. Geology of Portage County. *Report of the Geological Survey of Ohio* 3(1): 133–150.

Norling, Donald L. 1958. *Geology and Mineral Resources of Morgan County.* Ohio Division of Geological Survey Bulletin 56. Columbus, OH: Ohio Division of Geological Survey.

Norris, Stanley E., William P. Cross, and Richard P. Goldthwait. 1956 (1950). *The Water Resources of Greene County, Ohio.* Ohio Division of Water Bulletin 19. Columbus, OH: Division of Water.

Ollier, Cliff. 1969. *Weathering.* New York, NY: American Elsevier Publishing Co., Inc.

Orton, E. 1871. The geology of Highland County. *Geological Survey of Ohio Report of Progress in 1870*: 155–309. Columbus, OH: State of Ohio.

———. 1873. Geology of Clark County. *Report of the Geological Survey of Ohio* 1(1): 450–480.

———. 1874a. Geology of Greene County. *Report of the Geological Survey of Ohio* 2(1): 658–696.

———. 1874b. Geology of Pike County. *Report of the Geological Survey of Ohio* 2(1): 611–641.

———. 1878. Report on the geology of Franklin County. *Report of the Geological Survey of Ohio* 3(1): 596–646.

Overman, H. W. 1900. Fort Hill, Ohio. *Ohio Archaeological and Historical Publications* 1:260–264.

Parsley, Betty Jo. 1989. Natural oddities can be viewed in county. *The Messenger,* May 14: D-3.

Pettit, Lincoln. 1954. *Nelson Ledges: A Visitor's Guide.* East Lansing, MI: Michigan State University.

———. 1958. The effects of weathering and other changes at Nelson Ledges State Park. *The Ohio Journal of Science* 58(3): 182–186.

Pogue, J. E. 1911. The great Rainbow Natural Bridge of southern Utah. *National Geographic Magazine* 22: 1048–1056.

Portsmouth Area Recognition Society, The. 1986. *A History of Scioto County, Ohio*. Dallas, TX: Taylor Publishing Co.

Portsmouth Public Library. 2007. *Featured Photo: Devil's Tea Table. http:// yourppl.org/index.php?option=com_content&task=view&id=26&Itemid=41*. Accessed February 2, 2007.

Powell, Richard L. 1975. Theories of the development of karst topography. Pp. 217–242 *in* W. N. Melhorn and R. C. Flemal (eds.), *Theories of Landform Development: Publications in Geomorphology, Symposium Number 6*. Binghamton, NY: State University of New York.

Read, M. C. 1873. Geology of Geauga County. *Report of the Geological Survey of Ohio* 1(1): 520–533.

Rexroad, C. B., E. R. Branson, M. O. Smith, C. Summerson and A. J. Boucot. 1965. *The Silurian Formations of East-Central Kentucky and Adjacent Ohio*. Kentucky Geological Survey, Series 10 Bulletin 2. Lexington, KY: Kentucky Geological Survey.

Richard, Ben, and Mike Evers. 1990. *Ohio Academy of Science 1990 Geology Field Trip through Glen Helen*. Columbus, OH: The Ohio Academy of Science.

Roberts, David C. 1996. *A Field Guide to Geology: Eastern North America (The Peterson Field Guide Series)*. Boston, MA: Houghton-Mifflin Company.

Rogers, J. K. 1936. *Geology of Highland County*. Geological Survey of Ohio, Fourth Series Bulletin 38. Columbus, OH: Geological Survey of Ohio.

Rose, Bill (ed.). 1968. *Geological Aspects of the Maysville-Portsmouth Region, Southern Ohio and Northeastern Kentucky*. Lexington, KY: Kentucky Geological Survey.

Rosengreen, Theodore E. 1974. *Glacial Geology of Highland County, Ohio*. Ohio Division of Geological Survey Report of Investigations No. 92. Columbus, OH: Ohio Division of Geological Survey.

Scheidegger, Adrian E. 1970. *Theoretical Geomorphology*. Berlin, Germany: Springer-Verlag.

Shaver, Robert H., et al. 1978. *The Search for a Silurian Reef Model: Great Lakes Area*. Indiana Geological Survey Special Report 15. Bloomington, IN: Indiana Department of Natural Resources.

Slucher, E. R., E. M. Swinford, G. E. Larsen, G. A. Schumacher, D. L. Shrake, C. L. Rice, M. R. Caudill, and R. G. Rea. 2006. *Bedrock Geologic Map of Ohio*. Ohio Division of Geological Survey, Map BG-1, version 6.0. Columbus, OH: Ohio Division of Geological Survey.

Smith, Guy-Harold. 1935. The relative relief of Ohio. *The Geographical Review* 75: 272–284.

Snyder, T. A. 1988a. Ohio's natural bridges. *Division of Natural Areas Newsletter* 10(6): 2.

———. 1988b. Woodbury Wildlife Area Natural Bridge: Mini-karst in Coshocton County, Ohio. *Pholeos* 8(2): 3–4.

———. 1990. Ohio's natural bridges: A progress report. *Pholeos* 10(2): 7.

———. 2007. Natural arches in Ohio. *SPAN: Newsletter of the Natural Arch and Bridge Society* 19(4): 1–2.

———. n.d. *A Survey of the Natural Arches and Other Geological Features of Fort Hill State Memorial, Ohio.* Unpublished report on file with the Ohio Historical Society, Columbus, OH.

Stauffer, Clinton R. 1909. *The Middle Devonian of Ohio.* Geological Survey of Ohio, Fourth Series Bulletin 10. Columbus, OH: Geological Survey of Ohio.

Stauffer, Clinton R., George D. Hubbard, and J. A. Bownocker. 1911. *Geology of the Columbus Quadrangle.* Geological Survey of Ohio, Fourth Series Bulletin 14. Columbus, OH: Geological Survey of Ohio.

Stevens, Dale J., and J. Edward McCarrick. 1988. *The Arches of Arches National Park: A Comprehensive Study.* Orem, UT: Mainstay Publishing.

———. 1991. *The Arches of Arches National Park: Supplement.* Orem, UT: Mainstay Publishing.

Stauffer, Clinton R., George D. Hubbard and J. A. Bownocker. 1911. *Geology of the Columbus Quadrangle.* Geological Survey of Ohio, Fourth Series Bulletin 14. Columbus, OH: Geological Survey of Ohio.

Stout, Wilber. 1916. *Geology of Southern Ohio Including Jackson and Lawrence Counties and Parts of Pike, Scioto and Gallia.* Geological Survey of Ohio, Fourth Series Bulletin 20. Columbus, OH: Geological Survey of Ohio.

———. 1935. Rock resistance and peneplain expression. *The Journal of Geology* 43(8, Part 2): 1049–1062.

———. 1940. *Marl, Tufa Rock, Travertine, and Bog Ore in Ohio.* Geological Survey of Ohio, Fourth Series Bulletin 41. Columbus, OH: Geological Survey of Ohio.

———. 1941. *Dolomites and Limestones of Western Ohio.* Geological Survey of Ohio, Fourth Series Bulletin 42. Columbus, OH: Geological Survey of Ohio.

———. 1944a. Dolomites. *The Ohio Journal of Science* 44: 219–235.

———. 1944b. Sandstones and conglomerates in Ohio. *The Ohio Journal of Science* 44: 75–88.

Stout, Wilber, and G. F. Lamb. 1939. *Physiographic Features of Southeastern Ohio.* Geological Survey of Ohio, Reprint Series No. 1. Columbus, OH: Geological Survey of Ohio.

Stout, Wilber, and R. E. Lamborn. 1924. *Geology of Columbiana County.* Geological Survey of Ohio, Fourth Series Bulletin 28. Columbus, OH: Geological Survey of Ohio.

Stout, Wilber, Karl Ver Steeg, and G. F. Lamb. 1943. *Geology of Water in Ohio (A Basic Report).* Geological Survey of Ohio, Fourth Series Bulletin 44. Columbus, OH: Geological Survey of Ohio.

Stuckey, Ronald L. (ed.). 2003. *Linking Ohio Geology and Botany: Papers by Jane L. Forsyth.* Columbus, OH: RLS Creations.

Sturgeon, Myron T., et al. 1958. *The Geology and Mineral Resources of Athens County.* Geological Survey of Ohio, Fourth Series Bulletin 57. Columbus, OH: Geological Survey of Ohio.

Swinford, E. Mac. 1985. Geology of the Peebles Quadrangle, Adams County, Ohio. *The Ohio Journal of Science* 85(5): 218–230.

Termier, Henri, and Geneviève Termier. 1963. *Erosion and Sedimentation.* London, UK: D. Van Nostrand Co., Ltd.

Titamgim, R. Dirk. 1992. How are natural bridges formed? *Rocks & Minerals* 67: 410–415.

Totten, Stanley M. 1988. *Glacial Geology of Geauga County, Ohio.* Ohio Division of Geological Survey, Report of Investigation No. 140. Columbus, OH: Ohio Division of Geological Survey.

True, H. L. 1956. The Devil's Tea Table. *The Morgan County Herald.* January 19.

United States Geological Survey. 2007. *Geographic Names Information System Feature Query Results. http://geonames.usgs.gov/pls/gnispublic/.* Accessed February 21, 2007.

Van Tassel, C. S. 1901. *The Book of Ohio (Centennial Edition).* Bowling Green, OH: Self-published.

Ver Steeg, Karl. 1930a. Some features of Appalachian peneplanes. *Pan-American Geologist* 53: 359–364.

———. 1930b. Some features of Appalachian peneplanes. *Pan-American Geologist* 54: 17–28.

———. 1931. Erosion-surfaces of eastern Ohio. *Pan-American Geologist* 55:93–102, 181–192.

———. 1932. Nelson Ledges State Park. *The Ohio Journal of Science* 32(3): 177–193.

———. 1933. The state parks of Hocking County, Ohio. *The Ohio Journal of Science* 33(1): 19–36.

Ver Steeg, Karl, and George Yunck. 1935. Geography and geology of Kelley's Island. *The Ohio Journal of Science* 35(6): 421–433.

Vreeland, Robert H. 1976. *Nature's Bridges and Arches, Vol. 14 (Midwestern States).* Self-published.

———. 1994. *Nature's Bridges and Arches, Vol. 1 (General Information).* Self-published limited edition.

Walcott, Charles D. 1893. The Natural Bridge of Virginia. *National Geographic Magazine* 5: 59–62.

Wede, Henry. 2007. Arch hunting in the buckeye state. *SPAN: Newsletter of the Natural Arch and Bridge Society* 19(4): 8–11.

White, George W. 1979. *Extent of Till Sheets and Ice Margins in Northeastern Ohio.* Ohio Division of Geological Survey, Geological Note No. 6. Columbus, OH: Ohio Division of Geological Survey.

————. 1982. *Glacial Geology of Northeastern Ohio*. Ohio Division of Geological Survey, Bulletin 68. Columbus, OH: Ohio Division of Geological Survey.

White, George W., and Stanley M. Totten. 1985. *Glacial Geology of Columbiana County, Ohio*. Ohio Division of Geological Survey, Report of Investigations No. 129. Columbus, OH: Ohio Division of Geological Survey.

White, William B. (ed.). 1976. *Geology and Biology of Pennsylvania Caves*. Pennsylvania Bureau of Topographic and Geologic Survey, General Geology Report 66. Harrisburg, PA: Pennsylvania Geological Survey.

Wikimedia Foundation. 2006. *Tea table. http://en.wikipedia.org/wiki/Tea_table*. Accessed February 20, 2007.

Wilbur, Jay. 1993. *A Proposed Set of Standard Definitions for the Dimensions of Natural Arches (Preliminary Draft)*. The Natural Arch and Bridge Society. Unpublished manuscript in possession of the author.

————. 2004. *The Dimensions of Landscape Arch: Removing the Uncertainty. http://www.naturalarches.org/archinfo/index.htm*. Accessed November, 2006.

————. 2005. *Natural Arch Information. http://www.naturalarches.org/archinfo/index.htm*. Accessed November, 2006.

————. 2007. Arches beneath our feet. *SPAN: Newsletter of the Natural Arch and Bridge Society* 19(2): 10–12.

Winchell, N. H. 1874. Geology of Delaware County. *Report of the Geological Survey of Ohio* 2(1): 272–313.

Woodward, Herbert P. 1936. Natural Bridge and Natural Tunnel, Virginia. *Journal of Geology* 44: 604–616.

Index

1811 Arch 52, 199, 200, 398
Abutment *45*, 50, 96, 97, 104, 112, 115, 118, 134, 139, 140–142, 146, 176, 185, 197, 217, 229, 231, 236, 243, 244, 256, 259, 263, 264, 335, 336, 360
Adams County 9, 24, 30, 109, 111–113, 115, 116, 118, 120, 121, 124, 126, 127, 151, 297, 298, 300, 325, 344, 345, 394, 402
Adams, J. M. 163
Adelphi Creek 174
Alcove depth 44, 52, *54*
Algal stromatolites *290*
Allegheny Group 237, 239
Alluvium 174
American Columbo (*Frasera carolinensis*) 354
American Discovery Trail 373
Amherst Arch North 61, 206–208, 398
Ancestral Paint Creek 83, 84, 132
Andrews, E .B. 318
Andropogon 96
Annex Arch 196, 197
Ant lion 150
Anticline 21, 240, 241
Appalachian Basin 21
Appalachian Mountains 34, 172, 204, 221
Appalachian Plateau 5, 109, 250
Arbor Vitae (*Thuja occidentalis*) 118
Arch Rock xi, 243, 244, 383, 384, 400
Arch Trail (Miller Nature Sanctuary) 355, 356
Arches National Park 3, 13, 49, 339
Ash Cave 166, 167, 173
Ashtabula County 30
Athens County 242, 250, 312, 325, 326, 394, 402

Atmospheric erosion *16*, 56, 70, 76, 88, 94, 95, 97, 111, 112, 115, 129, 178, 211, 218, 341
Atmospheric weathering *16*, 68, 90, 171, 185, 196, 236, 243, 304
Bacon Flat 111
Baker Fork 37, 82–96, 99, 101, 103, 119, 372–377, **1**
Baker Fork Arch 94–96, 396, **1**
Baker Fork Gorge x, 83–85, 89, 91, 92, 94, 97, 98, 101, 102, 131, 289, 290, 338, 339, 372, 378
Balcony Natural Bridge 175, 176, 396, **9**
Baltic Pedestal Rock 327, 333
Base 1, 27, 42, *45–47*, 50, 51, 61–63, 74, 75, 78, 94, 104, 106, 107, 124, 149, 157, 198, 202, 245, *274–276*, 279, 283, 284, 293, 296, 298, 300, 310, 311, 315–319, 322, 326, 330, 331, 335, 341, 343–345
Base level 29, 92, 132, 135, 171, 174, 249
Bass Lake Natural Bridge 228, 229, 347, 396
Battle of Lake Erie 155
Beaver Creek Natural Bridge 6, 239–241
Bedding plane(s) 5, *33–36*, 42, 55, 61, 63–67, 69, 70, 74, 75, 78, 84, 87, 94, 108, 115, 122, 134, 138, 139, 141, 144, 145, 162, 178, 191, 218, 234, 238, 239, 255, 265, 275, 284, 292–294, 313, 315, 345, 395, 397, 399
Bedrock texture *67*, 68
Bedrock texture remnant arch 55, 56, 67–69, 162, 163, 196, 283, 395, 397, 399, 401
Bedrock texture remnant pillar 283, 290, 304

---

[1] In the pagination lists that appear in this index, italic face identifies pages that contain definitions, bold face identifies plate numbers, and normal face identifies pages that contain other references to the indexed entries.

Beech Flats 83, 84, 86, 92, 94, 101, 102, 374, 378
Bellefontaine Outlier 31, 32, 145
Belmont County 326
Berea Sandstone 206, 210, 377, 399
Big Cave Natural Bridge 99–101, 396
Big Natural Bridge 253, 256, 400
Big Pine Creek 301, 361, 362
Big Pine Creek Pillar 270, 300–302, 361–363
Big Rock 232
Big Run 231
Bigleaf Magnolia *(Magnolia macrophylla)* 235
Biostrome *290*
Birch, Hugh Taylor 388
Bird Run Standing Rock 327
Bisher Formation 151, 395
Black Hand Sandstone 25, 32, 42, 165–168, 170, 172, 173, 176, 180, 181, 186, 191, 192, 196, 197, 199, 221, 300, 304, 305, 397, 399, 403, **10, 11**
Black River 62
Blind arch 160, *205*
Blocked Arch 118, 120, 121, 394
Botryoids *90*, 97, 104, 108
Boundary Arch 44, 63, 85, 86, 396
Bouquet, Colonel Henry 241
Box canyons 166, 173
Brachiopods 159
Brady, Samuel 332
Braun, E. Lucy 89, 96
Breccias *83*
Breeched alcove arch *56*, 62, 66, 103, 201, 208, 230, 242, 263, 395, 397, 399, 401
Bridge Creek 90, 91, 93, 94, 376
Brineinger Hollow Natural Bridge 180
Buckeye Trail 362, 373
Buffalo Clover *(Trifolium reflexum)* 232
Buffalo Sandstone 329
Bundle Run Arch 107, 121–123, 394
Burnett Run 257
Butterfly Weed *(Asclepius tuberose)* 149
Buzzardroost Rock 127

Cadiz Standing Stone 329, 330, 346
Calcareous rock 26, 78, 150
Calcium carbonate 18, 26, 78, 79, 90, 108, 130, 131, 152, 196, 261, 264, 284, 286, 287
Caldwell, J. A. 326
Camp Christopher Natural Bridge 226–228, 388

Campbell Hill 29
Canada Yew *(Taxus canadensis)* 96
Canters Cave Natural Bridge 268
Carbonate 33, 71, 81, 109
Cascade Park 62
Case hardening *184*
Castalia Quarry Reserve 60, 79, 80
Castlegate Arch xiii, 12, 127–129, 394
Catawba Point 341
Cats Den Natural Bridge 63, 222–224, 226–228, 254, 396
Cats Den Road 222, 226
Cave collapse 2, 83, 126
Cave Collapse Arch *71*, 72, 395, 397, 399
Cedar Falls 173
Cedar Fork Arch 67, 68, 116–118, 325, 350, 360, 394
Cedar Fork Standing Stone 118, 325
Cedarville Dolomite 82, 137–139, 142–146, 149, 292, 294, 295, **18**
Cedarville-Springfield-Euphemia dolomites 141, 144
Cemented glacial gravel 261, 263, 265, **17**
Cenozoic Era 18, 19
Champaign County 11, 159
Chapel Ridge Natural Bridge xi, 181
Chemical weathering *15*, 16, 25, 26, 35, 62, 284
Chief Leatherlips 163, 164, 389
Chimney 124, 271, *276*, 281–283, 307, 309, 311, 326, 403
Chimney Rock 232, 271, *276*–281, 307, 308, 309, 310, 338, 402
Chimney Rock Hollow 326
Cincinnati Arch *20*–22
Cincinnati River 174
Civil War 155, 191, 244, 385
Clark County 139, 394
Clark, George Rogers 139
Clark, William 294
Clastic rock *17*, 18, 22, 33
Clearance *(see* individual arches for a discussion of their clearance.) 12, 44, *46*, 47, 51, 52
Cleland, Herdman F. 3, 40
Clifton Gorge 75, 77, 78, 138, 142, 145, 148, 253
Coastal arch 56, 153
Cobble Arch 263–265, 398, *17*
Col *84*, 91, 92, 101, 374
Colonel Drake 246
Colorado Plateau 2, 40, 181

Columbiana County 239
Columbus Limestone159, 160, 339, 397
Columbus, Hocking Valley and Toledo Railway 168
Common Polypody (*Polypodium vulgare*) 184, 185
Coneflower (*Ratibida pinnata*) 149, 386
Conemaugh Group 243, 329, 401
Conglomerate 16, 17, 22, 24, 26, 67, 165, 220, 221, 223, 224–226, 229, 231, 234, 261, 263–266, 307–309, 389
Conkle, W. J. 183
Conkles Hollow Arch 7, 183, 185, 396
Conkles Hollow State Nature Preserve 183, 184
Continental divide 29, 138, 207
Cooke, Jay 155, 385
Coshocton County 237, 326, 370
Council Rock 161, 162
Covered Arch 114–116, 3
Covered Bridge Scenic Byway 379
Crane Hollow 181, 199, 200
Crawl Arch 118–121, 394
Crinoidal hash 153
Cross-bedding 67, 165, 184, 196, 197, 206, 211, 221, 231, 315
Cute Tower 300, 301, 402
Cuyahoga Formation 165
Cuyahoga River 332, 333

Dale Tea Table 313–315, 402
David Road Natural Bridge 262, 263, 346
Davis, David 88
Davis Cave 88
Dayton Power and Light Company 387
Dayton-Cincinnati Pike 268
Deck 48, 108, 206, 247, 254, 337
Deck height 44, 53
Deep Stage Drainage 174
Deer Trail (Fort Hill State Memorial) 91, 373, 377
Delaware County 161, 163, 396
Delaware Limestone 162, 163, 397
Delta 165, 209, 221
Deposition arch 78, 79
Deposition pillar 286
Devils Bathtub 181, 183, 364
Devils Tea Table (Morgan County), The 6, 316–322, 339, 346, 391, 402
Devils Tea Table 281, 313, 316, 327, 332, 334
Devonian Period 19, 21, 32, 131, 159, 165, 376

Differential cementation 166, 218
Differential solution 33, 69, 389
Differential weathering 199, 220, 315
Dimetrodon 249
Dolomite 16, 18, 22, 61, 67, 71, 74, 80, *81*, 82 , 83, 85, 86, 94, 97, 101, 104, 106, 107, 109, 111, 118, 121–123, 127, 128, 135, 144, 146, 148, 157, 264, 290, 292, 293, 298, 300, 341, 342, 355, 369, 373, 376, 388
Double Dome Cave 99
Dublin Arch 159–161, 396
Dunbar Tea Table 323, 402
Dunkard Group 250, 401, 403

Eagle Rock Arch 266, 364
Early Arch 59, 61, 167, 200, 201, 398
Eastern Deciduous Forest 261
Eastern Hemlock (*Tsuga canadensis*) 228, 235
Eastern Wood Rat (*Neotoma floridana*) 298
Easton Hill 377
Edaphosaurus 249
Edge of Appalachia Preserve xi, 127, 297, 298, 300
Elizabeth L. Evans Outdoor Education Center Canters Cave 4-H Camp, Inc. 268
Elk Fork 243
Enlarged joint and bedding planes arch 55, 56, 63, 65, 66, 234, 395, 397, 399
Enlarged vertical crevice arch *56*, 64
Entrance (*see* individual arches for a discussion of their entrances.) ii, *44*, 45–48, 52–54, 57, 63, **6, 11, 12**
Erie County MetroParks 80
Erie Islands 21, 156, 276, 339
Erosion 2, 15, 16, 18, 19, 21, 23, 25, 28, 32, 34, 36, 38, 40, 43, 55, 59, 62, 63, 66–68, 70, 72, 74, 78, 82, 83, 84, 86–88, 92, 95, 96, 100, 102, 109, 111, 113, 115, 116, 119, 132, 137–139, 144, 146, 156, 165, 189, 196, 210, 225–226, 227, 238, 242, 244, 248, 249, 252, 257, 261, 272, 282, 284, 290, 293, 296, 300, 304, 311, 319, 320, 323, 335, 341, 344
Erosion remnant 269, 274, 275, 276, 317
Euphemia Dolomite 139, 141, 144, 293, 296

Fairfield County 168, 171, 277, 396
Falls Trail (Miller Nature Sanctuary) 105, 355, 356
Faults 33
Fayette Township 268

Fern Pillar 261, 324, 379, 382, 402
Fin 48, 60, 61, 128, 129, 153, 154, 197, 206, 243
Findlay Arch 21
Flat Iron Rock 199, 303–306, 402, **20**
Flowerpot Rock 341–344
Flowstone 78, 90, 98, 104, 108, 115, 389
Flushing Escarpment 32
Fluting 238, 239
Foerste, A. F. 262, 263
Fool's gold 221
Fort Ancient 378
Fort Hill State Memorial x, 72–74, 82, 85, 92, 96, 99, 131, 151, 372, 374, 375
Fort Hill Tea Table 95, 289, 290, 402
Fort Trail (Fort Hill State Memorial) 377
Fossils 67, 159, 220, 249, 290
Foster, Charlie xi
Foster, James A 268
Franklin County 66, 159, 396
Franz Theodore Stone Laboratory 384
Front *48*, 53, 54, 57, 85, 87, 89, 94, 104, 108, 111, 115, 117, 123, 124, 148, 150, 170, 180, 182, 191, 192, 217, 218, 220, 229, 234, 251, 255, 256, 257–259, 263, 293, 311, 375, 389
Frost Cave 150
Fuller, Nate xi, 154

Gateway, The 179, 337, 338
Geauga County 222, 228, 254, 266, 327
George Rogers Clark Park 139
Gibraltar Island 153–155, 385
Gilmer, Francis William 2
Glacial derangement 130, 173
Glacial diversion 27, 29, 32, 37, 38, 84, 240
Glacial erratic 227, 264, 274
Glacial till 78, 103, 156, 160, 162
Glen Helen Ecology Institute 387
Goblet Rock 326, 327
Gold Hunters Cave 221
Gorge Trail (Fort Hill State Memorial) 87, 96, 364, 373–377
Goss Fork 244, 245
GPS 7, 9
Gravity arch *42*, 72, 73–77, 156, 221, 222, 301, 362, 376
Gravity pillar *274*, 284, 285
Gray-headed Coneflower (*Ratibida pinnata*) 149
Great Cave 380, 381
Great Lakes-Saint Lawrence Watershed 29

Great Laurel (*Rhododendron maximum*) 171, 172
Great Miami River 216
Great Road, The 241
Greene County 6, 78, 142, 291, 294, 315, 388, 396, 402
Greenville Falls Arch 139, 146, 147, 150, 264, 385, 387, 399, **5**
Greenfield Dolomite 34, 82, 83, 96, 131–136, 399
Greenville Falls State Nature Preserve 150, 385, 386
Groundwater 25, 62, 67, 71, 82, 86, 90, 98, 111, 126, 144, 148, 149, 162, 163, 178, 194, 202, 221, 223, 224, 234, 255, 261, 264, 282–284, 337, 387
Groveport River 174
Guernsey County 327, 329
Gulf of Mexico 29, 165
Gulf of Queer Creek, The 175
Gunpowder 201, 202, 204

Hagley Hollow Arch 6, 186, 187, 398, **10**
Hamilton County 29
Hanging Rock region 244
Hansen, Michael C. x, 6, 165, 230
Harebell (*Campanula rotundifolia*) 149
Harrison County 6, 329, 330
Harrison, Joseph T. 329, 330
Hayden Falls 66
Headland 91, 157, 207, 210
Headward erosion 2, 60, 230, 399
Heart-leaved Plantain (*Plantago cordata*) 118
Hermit Cave 88
Hewett Fork 312, 326
Hidden Arch 101, 102, 339, 396
Highland County 24, 37, 72, 83, 85, 86, 89, 94, 96, 98, 99, 101, 103, 104, 107, 116, 132, 289, 372, 396, 402
Hintz Hollow Arch 171, 172, 396
Hockhocking 167
Hockhocking Island 252
Hocking County 24, 88, 165, 167, 168, 175, 176, 180, 181, 183, 184, 186, 187, 189, 191, 196, 197, 199, 200, 201, 203, 204, 242, 266, 270, 300, 302, 304, 337, 396, 398, 402
Hocking County Balanced Rock 302, 303, 304, 305, **20**
Hocking Hills State Park 167, 173, 182, 191, 267, 362–364

Hocking River 167, 168, 170–172, 325, 351, 352
Hocking State Forest xi, 173, 175, 300, 361
Hocking Valley Birding Trail 351
Hocking Valley Canal 168
Hockingport Sandstone Lentil 250
Hole-in-the-Wall Arch 52, 233, 234, 237, 398
Holocene Epoch 37
Honeycomb texture 220
Hoodoos 271
Hopewell Culture National Historical Park 378
Horse Thief Cave 249
Howe, Henry 4–6, 248, 316, 317, 319, 320
Howes, Mark vii, xii, 183, 184
Hunt, Ken 294
Hydration 26
Hydrostatic pressure 239

Ice Age 19
Igneous rock 17, 18, 163, 264
Illinoian glaciation 37, 68, 97, 101, 127, 129, 170
Illinois Basin 21
Independence (Irish Run) Natural Bridge 5
Indian Council Chamber 161, 163, 389
Indian Watch Tower 330
Irish Run Natural Bridge 6, 258–260, 324, 379–383, 400
Iron compounds 165, 196, 221
Iron pyrite 221
Ironstone 220
Ironton Boulder 330, 331

Jackson County 233, 234, 236, 268, 332, 398
Jarnigan Knob 90, 91, 374
Jawbone Arch 53
Jawbone Arch — North 107, 396
Jawbone Arch — South 107, 396
Jefferson County 11, 249, 326
Jefferson, Thomas 2, 55
John Bryan State Park 76, 142, 285, 294, 388
John Bryans Window 140, 141, 294–296, 402, 19
Johnson, Jeff xi, 180, 203
Joint control 87, 88, 154, 160, 170, 229, 258, 304
Joints 34, 35, 42, 56, 57, 63, 65, 66, 86–88, 93, 94, 103, 119, 120, 123, 129, 135, 158, 176, 181, 185, 192, 195, 197, 205, 225, 234, 248, 266, 282, 293, 304, 315, 341, 13
Jones, Edwin A. 235

Jughandle Falls 78

Kame 293, 344
Kankakee Arch 21
Kansan glaciation 37
Karst 26, 83, 109, 150, 160, 237, 238
Kellys Island 156
Kent Standing Stone 332, 333
Kentucky vii, 2, 3, 5, 13, 43, 109, 130, 133, 212, 244, 280, 6
Keyhole Bridge 89
Keyhole, The 6, 43, 63, 64, 89–94, 372, 375, 376, 396
Kind, Thomas C. 5

Ladd (Big) Natural Bridge vii, 5, 6, 36, 65, 253–256, 335–337, 350, 360, 400, 15
Lake Erie xi, 6, 17, 29, 31, 37, 145, 153–157, 159, 207–209, 214, 276, 339–341, 344, 385, 7
Lake Katharine Arch 234–236, 398
Lake Katharine State Nature Preserve 234
Lake Lundy 208
Lake Maumee 208
Lake Superior 244
Lake Warren 208
Lake Whittlesey 207–210
Large (Ladd) Natural Bridge 5
Large-flowered Trillium (Trillium grandiflorum) 373, 376
Laurel Run 191
Laurelville Creek 173
Lawrence County 268, 330, 331
Leatherlips Arch 161, 163, 164, 389, 396
Ledges 75, 144, 166, 220, 223, 226, 248, 249, 255, 389
Leg 277, 279, 280, 284, 290, 291, 293–296, 300–302, 310–312, 315, 319, 328, 330, 331, 338, 340, 344, 18, 19
LeMarin Arch 155–158, 398
Liberty Natural Bridge 244–246, 400
Liberty Township 246
Licking River 165
Limestone Savory (Calamintha arkansana) 149
Lintel 44, 45, 46, 48, 50–53, 55–57, 78, 80, 85–88, 90, 91, 94, 96, 97, 100–102, 104, 105, 108, 113, 115, 117, 123, 127, 133, 139, 141–144, 148–150, 152, 154, 160, 166, 170, 175, 176, 181, 182, 198, 206, 217, 218, 225, 227–229, 231, 234, 235, 240, 244, 245, 247–249, 251, 254–260, 263, 324, 335, 2, 13

Lintel breadth 53
Lintel depth 53
Lintel thickness 44, 53
Lintel width 44, 48, 53
Lions Den Arch 150, 151, 398
Little Chimney Rock 308–310, 402
Little Miami River 138, 142, 145, 294
Little Muskingum River 379
Little Natural Bridge 256–258, 400, **16**
Logan County 29, 145
Logan Formation 210, 216
Lorain County 165, 206, 398
Losantiville 253
Low Arch 142–145, 396
Lower Falls (Hocking Hills State Park) 364
Lower Freeport Sandstone 239, 312, 403
Lucas Run Natural Bridge 57, 247, 248, **14**
Luther, Warren Phillips 266, 327

M Arch 197–199, 304, 398
Macksburg Field 246
Mad River Gorge 139, 142, 145, 148
Madison Township 332
Mahoning Trail 332
Maidenhair Spleenwort (*Asplenium trichomanes*) 256
Manchester–Old Kentucky River system 129
Mansfield Natural Bridge (IN) 346
Marblehead Peninsula xi, 21, 156, 157, 276, 339, 341, 342
Marietta and Cincinnati Railroad Company 242
Marietta Unit (Wayne National Forest) 379
Marl *95*, 96, 290
Massies Creek 138
Mattress Arch 118, 119, 360, 394
McKitterick, James J. 235
Meander arch 61, 135
Mechanical weathering 15, 16, 27, 272
Meigs County 249
Merkles Cave 266
Merkles Natural Tunnel 266
Mesozoic Era 18, 249
Metamorphic rock *17*, 18, 163, 264
Miami County 146, 263, 265, 398
Miami County Park District, The 385, 386
Miami Township 268
Michigan Basin 21
Middle Ridge Road 209
Miller, Arthur M. 2
Miller Arch 39, 40, 70, 104–106, 350, 356, 396, **2**

Miller Natural Bridge 39, 57, 103, 104, 108, 350, 355, 396
Miller Nature Sanctuary 104–106, 354
Mineral Arch 6, 242, 243, 271, 312, 394
Mineral Tea Table 243, 271, 312, 313, 402
Mississippian Period 19, 21, 25, 29, 165, 206, 210, 220, 221, 300, 377, 401
Mississippian sandstone 210, 401
Mohican State Park 25
Monitor 244
Monongahela Group 244, 247, 315
Monroe County 29
Monroe Limestone 150, 399
Montgomery County 262, 268
Montgomery County Arch 268
Moorefield Township 330
Morel Tea Table 310, 311, 402
Morgan County 6, 246, 247, 313, 316, 339, 391, 398, 402
Mount Pleasant 277, 278
Mountain Laurel (*Kalmia latifolia*) 256
Mountain Lion (*Felis concolor*) 226
Muddy Crack Arch 204, 205, 398
Munson Township 266
Murphy, James L. viii, x, 5, 7, 239, 252, 326–334
Mushroom rocks 271
Muskingum River 316, 321, 346
Mustapha Natural Bridge 6, 250–252, 394

Narrow-leaved Blue-curls (*Trichostema dichotomum* var. *lineare*) 231
Natural Arch and Bridge Society, The 3, 6
Natural arch ii, ix, x, 1, 5, 12, 20, 36, 39, 40, *41*, 42, 43, 46–50, 55–58, 60, 72, 75, 130, 132, 135, 142, 168, 183, 191, 214, 252, 263, 284, 337, 338, 351, 364
Natural Areas Act of 1970 350
Natural Bridge (VA), The 2, 54
Natural Bridge State Park (KY) vii
Natural bridge viii, 4, 5, 10, 39, 40, 41, 43, 48, 55, 63, 66, 78, 87, 90, 100, 104, 168, 169, 171, 179, 182, 188, 189, 223, 224–227, 229, 230, 237, 239–241, 245–248, 251, 252, 254–260, 263, 268, 335–337, 346, 347, 351–353, 360, 370, 371, 380, 382, 388, 289
Natural pillar ix, 5, 8, 9, 12, 16, 17, 28, 32, 36, 37, 118, 143, 156, 269, 271–273, 275, 277, 279, 280, 281–285, 289–293, 294–296, 297–300, 303, 304, 307, 312, 313, 316–317, 319–321, 323–328, 330, 333, 335, 339, 341, 343, 344, 372

Natural tunnel ii, 43, 89, 90, 91, 94, 191, 192, 194–196, 266, 364
Natural Y Arch 6, 86–88, 372, 375, 396
Nature Conservancy, The xi, 127, 297, 325
Naylor, James Ball 317, 318
Needles Eye 6, 153–156, 159, 384, 385, 398, 7
Neff Park 291
Nelson-Kennedy Ledges State Park 221, 222
Newark River 174
Niagara Escarpment 30, 37, 137, 142, 145
Noble County Rocks 331, 332
Nodding Wild Onion (*Allium cernuum*) 149
North Country National Scenic Trail 373, 380
North Fork of Little Beaver Creek 239–241
North Sugar Rock Arch 157, 158, 398
North Township 330

Off-loading *34*, 63
Ohio and Erie Canal 332
Ohio Archaeological and Historical Society 6
Ohio Brush Creek 82, 126, 127, 297, 298, 300
Ohio Cave Survey xi, 109, 111, 126, 135, 150
Ohio Caverns 281, 285
Ohio Department of Natural Resources (Ohio DNR) 5, 19, 20, 22, 23, 308, 313, 342, 350, 361–363, 367, 368, 387
Ohio Department of Transportation 9
Ohio Division of Forestry (Ohio DOF) xi, 361
Ohio Division of Geological Survey (Ohio DGS) xi, 5–7, 154, 187, 230, 328
Ohio Division of Natural Areas and Preserves (Ohio DNAP) vii, viii, xi, xii, 5–7, 9, 12, 104, 118, 180, 183, 184, 201, 203, 216, 235, 247, 256, 331, 333 350, 351, 386
Ohio Division of Parks and Recreation (Ohio DPR) 362, 363
Ohio Division of Wildlife (Ohio DOW) 367, 368, 371
Ohio Historical Society x, 6, 82, 372
Ohio Natural Arch Survey x, 1, 128, 183
Ohio River 21, 29, 31, 128, 129, 139, 151, 159, 165, 174, 204, 210, 212, 216, 219, 249–252, 267, 297, 344, 356–358, 379, **6**
Ohio Sea Grant College Program 385
Ohio Shale 91, 93, 94, 131, 159, 376
Ohio-Mississippi River watershed 29
Ohioview Arch 151–153, 394, **6**
Old Kentucky River 129, 216
Old Mans Cave 88, 173, 181, 182, 190, 191, 266, 363, 364

Old Mans Pantry 189–191, 267, 362, 364, 398
Old Womans Kitchen — East Arch 124, 125, 394
Old Womans Kitchen — West Arch 124, 125, 394
Old Womans Kitchen 124, 125
Olentangy Caverns Arch 69, 161–163, 389, 396
Opened bedding planes arch 55, 61, 69, 399
Opening (*see* individual arches for discussions of their openings.) 12, 39, 41, 42, *44*, 45, 46, 47, 49, 50–53, 57, 58, 63, 65, 66, 70, 71, **1, 4**
Ordovician Period 19, 21, 81, 137, 289
Orton, Edward 309
Ortt, Marilyn xi, 244, 247
Ottawa County 153, 155, 157, 341, 398
Outwash 145, 171
Oxidation 26

Paint Creek 102, 132, 133, 368, 378
Paint Creek Reservoir 368
Paint Creek State Park 368
Paint Creek Lake Wildlife Area 34, 35, 367
Paleokarst 83, 96, 106, 111, 369
Paleozoic Era 18, 19
Passage, The 98, 99, 396
Patterson, John 253
Patterson, Robert 252
Peach Mountain 116
Pebble lens 165, 234
Pebbles 165, 220, 221, 224, 233, 234
Peckuwe 139, 214
Pedestal Rock 331, 339
Peebles Dolomite 35, 70, 82–84, 91, 93–96, 99–102, 106, 107, 109, 111, 113, 115, 116, 119, 121, 124, 126, 128–131, 133, 134, 137, 138, 290, 297, 298, 300, 344, 369, 373–376, 395, 397, 401, 403
Peebles, J. C. 359
Peephole Arch 236
Pennsylvanian Period 19, 21, 220, 237, 249, 307
Penthouse, The 219, 400
Permian Period 19, 21, 249, 256, 258, 261, 323, 324
Perry, Oliver Hazard 154, **7**
Pholeos 135, 137, 266, 327
Pickaway County 174
Pickawillany Trail 214
Pike Arch 12, 60, 230–234, 347, 398
Pike County 12, 150, 230, 232, 276, 307, 309, 310, 332, 398, 402

Pillar width 278, *279*
Pillared alcove *52*, 61, 67, 151, 156, 166, 177, 178, 189, 190, 199–202, 219, 267, 354, 364, 395, 399
Pit 63, 72, 212, 222–230, 252, 346, 347
Platform *274*, 279, 293, 297, 302–304, 306, 309, 312, 315
Pleistocene Epoch 17, 19, 27, 36, 38, 82, 129, 174, 193, 194, 216, 222, 241, 249, 261, 399
Polypody Arch 184–186, 396
Pomeroy River 174
Pompeys Pillar 6, 141, 142, 291–294, 315, 387, 388, 402, **18**
Porosity 33, 35, 67, 81
Portage County 221, 332
Portage Escarpment 29, 31
Portsmouth River 129, 216
Potassium nitrate 201, 202, 204
Pothole 62, 181, 182, 364
Pottsville Group (?) Sandstone 395
Powder mill 201
Precambrian Era 18
Pre-glacial divide 29
Prickly Pear (*Opuntia humifusa*) 231
Prider Road Pillars 329
Prow Tower 298–300, 402
Put-in-Bay Dolomite 153, 154, 399
Putnam Hill Limestone 237, 238, 397

Quartz 220, 221, 233
Quaternary Period 19, 37, 175, 261
Queer Creek 172–175

Ramey, Ralph xi, 293
Rattlesnake Creek 368
Raven Rock Arch 51, 210–214, 216, 217, 337, 350, 357–359, 400, **13**
Reach (*see* individual arches for a discussion of their reach.) 45, *46*, 47
Recessional moraine 108, 148
Red River Gorge (KY) 13, 60
Reindeer Moss (*Cladonia* spp.) 231
Resurrection Fern (*Pleopeltis polypodioides*) 118
Rim Trail (Hocking Hills State Park) 365, 367
Riverbend Arch 9, 109–111, 394
Roadside Arch 111, 112, 394
Roadside Cave 111, 113, 114
Rock Run 234, 235
Rockbridge State Nature Preserve 168, 351
Rockbridge vii, 4, 6, 39, 401 168–171, 350–353, 361, 392, 396, **8**

Rockgrin Arch 12, 216, 217, 219, 350, 360, 400
Rock House ii, 6, 40, 43, 52, 53, 173, 177–179, 191–196, 362, 364–367, 398, **11, 12**
Rocky Fork 39, 72, 73, 82, 102, 103, 105, 106, 132, 354–356
Rocky Fork Gorge State Nature Preserve 354
Rocky Fork Needles Eye 72, 73
Rocky Mountains 306
Rome Township 325, 331
Rosebay (*Rhododendron maximum*) 171
Ross County 130, 132, 135, 398, 400
Round-leaved Catchfly (*Silene rotundifolia*) 231

Salt Fork 174
Saltpetre Cave 201, 203
Saltpetre Cave Arch 201–203, 350, 353, 354
Saltpetre Cave State Nature Preserve 201, 353
Sand Hill 327
Sapping 66, *176*, 229, 239
Scio Stone 330
Scioto Brush Creek Arch 113–115, 394, **3**
Scioto County 165, 210, 215–217, 219, 267, 332, 400
Scioto County Devils Tea Table 332, 334
Scioto River 66, 159, 174, 214, 216, 378
Scouring-rush (*Equisetum hyemale*) 356
Sea arch 56
Sea stacks 271, 275, 276
Sedimentary grain 283
Seven Caves, The 103, 106
Sewickley Sandstone 244
Shadow Rock 325
Sharon Formation 32, 75, 220–222, 226, 228–230, 232–234, 266, 307, 310, 327, 397, 399, 401, 403
Shawnee State Park 210
Shoemaker State Nature Preserve 116, 360
Shooting Star (*Dodecatheon meadia*) 354
Shrubby Cinqufoil (*Dasiphora fruiticosa*) 149
Siltstone 165
Silurian Period 19, 21, 32, 81, 82, 103, 106, 132, 137, 142, 150, 151, 153, 159, 264, 289, 292, 376
Simmons Run 237
Sinkhole 72, 106, 126, 127, 136–137, 163, 265, 266, 369, 371
Skull Cave Arch 130, 131, 135–137, 369, 371, 398
Skylight (*see* individual arches for discussions of their skylights.) *44*, 45, 53–55, 57, 62, 63, 65–67, **1, 3, 5, 17**

Skylight length 44, *53*
Skylight width 44, *53*
Slab Arch 74
Slide Arch 217, 218, 350, 360, 400
Slip-off slope 260
Slump block 42, 72, 74–78, 141, 153, 171, 245, 285, 315, 322, 338, 355, 362
Small Roadside Arch 111, 112–114, 394
Small Standing Stone 326
Smaller Roadside Arch 111, 113, 114, 394
Snow Trillium (*Trillium nivale*) 354
South Bass Island 153, 156, 385
South Sugar Rock Arch 157–159, 398
Span (*see* individual arches for a discussion of their span.) 12, 16, 43–45, *46, 47,* 48, 50–53, 57, 63, 78
SPAN: Newsletter of the Natural Arch and Bridge Society 3, 6
Spencer Creek 326
Spring (water feature) 36, 78, 146–149, 286
Spring Creek Arch 96, 97, 101, 116, 151, 372, 376, 396
Springfield Dolomite 137–142, 144–146, 149, 214, 293–296, 395, 397, 403, **18**
Stalagmite 284, 285
Standing stone 277, 278, 324, 332, 382
Standingstone Indian Trail 329
Stevens and McCarrick 49
Stillwater State Scenic River 150, 386
Stone Lab 155, 385
Stout, Wilber 174, 221, 309
Stream erosion 70
Sugar Rock 157, 158
Sullivantia (*Sullivantia sullivantii*) 118, 186
Sulphur Creek Gravity Arch 72, 73, 376
Sunfish Creek 150
Surprise Arch 203, 350, 354, 398

Table (*see* individual pillars for a discussion of their table.) 277, 279, 280, 284, **18**
Table Rock 339–341
Tea table 271, *276,* 277, 280–284, 290, 292–294, 296–298, 300, 303, 304, 311, 312, 315, 316, 319, 320, 322, 328, 330, 333, 339, 344, 345, **19**
Teakettle Rock 297, 402
Teapot effect 284, 290
Teays River 129, 170–173, 216, 298
Tecumseh Arch 12, 70, 139–142, 145, 294, 394
Tecumsehs Shooting Gallery 139
Tension dome *58,* 59, 60, 135, 198, 200, 242, 337

Tertiary Period 38, 163, 171, 249
Three Hole Arch 176, 177, 179, 337, 338, 398
Tiffin Arch 71, 126, 127, 394
Till *37,* 78, 103, 148, 156, 160, 163, 261
Titusville Till 240
Tolle Hill 116
Top *274,* 276, 279, 283, 284, 294, 298, 313, 319, 320, 323, 326, 328–331, 333, 338, 341
Topographic relief 16, 21, 28, 32, 264
Toppled tea table 344, 345
Trautwine, John C. 2
Travertine mound 78
Trenton Limestone 20, 21
Trimmer Arch 61, 106, 131–136, 367, 369, 370, 398, **4**
True, H. L., Dr. 316, 317, 319
Tuliptree Trail (Miller Nature Sanctuary) 356
Twain, Mark 281
Twin Falls 327
Tymochtee Dolomite 156, 159, 399

Umbrella Magnolia (*Magnolia tripetala*) 235
Unconformities *18,* 19
Unger Hollow Natural Bridge 187–189, 398
Uniontown Sandstone 326
Upper Falls (Hocking Hills State Park) 181, 190, 191, 364
US Army Corps of Engineers 346, 368
US Forest Service 379
US Geological Survey 92, 197, 280

Vertical crevice enlargement 395, 397, 399, 401
Vertical Dome Arch 130–132, 400
Vinton County 243, 400
Vinton Furnace 244, 384
Vinton Furnace Experimental Forest xi, 244, 383, 384
Virginia Bluebells (*Mertensia virginica*) 356
Vreeland, Robert H. 3, 5, 6, 13, 14, 48, 187
Vugs *83*

Walking Fern (*Camptosorus rhizophyllus*) 256
Wand-lily (*Zigadenus elegans*) 149
War of 1812 154, **7**
Warren County 378
Warren Township 326
Warriors Path 214
Washington County 244, 246, 247, 253, 256, 258, 313, 323, 324, 333, 335, 402
Washington, George 55
Waterloo Wildlife Experiment Station 326
Watershed 32, 90, 257, 361

Weathering (*see* also: atmospheric weathering, atmospheric erosion, differential weathering, chemical weathering, mechanical weathering) 16, 26–28, 33, 36, 37, 57, 62, 66, 67, 70, 74, 83, 129, 130, 141, 148, 160, 165, 166, 170, 171, 176, 180, 184, 189, 191, 192, 199, 220, 221, 224, 229, 236, 255, 269, 281–284, 296, 298, 309, 310, 335, 337, 339

Wede, Henry x, 6

Wells, L. M. 163

West Branch of Wolf Creek 315

Whan, Pete xi, 297, 298

Wheeling Township 328

White Cedar (*Thuja occidentalis*) 118

Whites Gulch 232

Wild Columbine (*Aquilegia canadensis*) 354

Wills Creek 327

Wisconsinan glaciation 37, 130, 138, 142, 145, 148, 226, 240, 264, 293

Wittenberg University Speleological Society 135

Wood Rat Tower 298–300, 402

Woodbury Natural Bridge 237, 238, 267, 270, 396

Woodbury Wildlife Area 237, 370

Wyandot Indians 163

Yellow Spring 286, 292, 388

Yellow Springs Creek 293, 388